Laboratory Manual

to accompany

Adventures in Chemistry

Laboratory Manual

to accompany

Adventures in Chemistry

Julie Millard, *Colby College*

Rebecca J. Rowe, *Colby College*

Mark Blaser, *Shasta College*

HOUGHTON MIFFLIN COMPANY Boston New York

Vice President and Executive Publisher: George Hoffman
Vice President and Publisher: Charles Hartford
Senior Marketing Manager: Laura McGinn
Senior Development Editor and Director of Development: Bess Deck
Assistant Editor: Amy Galvin
Editorial Associate: Henry Cheek
Senior Project Editor: Charline Lake
Marketing Assistant: Kris Bishop

Printed in the U.S.A.

ISBN-13: 978-0-618-37664-3
ISBN-10: 0-618-37664-X

1 2 3 4 5 6 7 8 9-CRS-10 09 08 07 06

TABLE OF CONTENTS

PREFACE

To the Student

Welcome to the *Laboratory Manual* to accompany *Adventures in Chemistry*! This lab manual was designed to follow the text by presenting experiments with real-life applications. In the first experiment, you will be formally introduced to UniChem International, a fictional major chemical conglomerate that includes a Forensics Department, a Chemical Analysis Department, a Criminal Division, and Synthesis and Biochemistry Laboratories. Each week you will be assigned to a different sector and given tasks specific to each department. This may include, among other procedures, forensic analysis, synthesis, and qualitative analysis

Although UniChem International is fictional, the assignments you complete will give you an idea of what really happens in these different chemical environments. After this course is completed, you may never work in another chemistry laboratory—our goal is to help you to understand why chemists find working in the laboratory so fascinating, interesting, and fun.

For some of you, this may be your very first time in a chemistry laboratory, but don't panic! You are about to experience one of the most exciting scientific adventures possible. Look around the laboratory—you may see equipment and glassware that you have never seen before. By the end of the semester, you will have used most of these and will have become very familiar with their uses.

The experiments in this lab manual were written with two goals in mind: first, to reinforce the concepts that you are learning in the lecture portion of this course and, second, to introduce you to the real-world applications of chemistry. Some of the material you will be using in the lab may include common household items such as vinegar (acetic acid), baking soda (sodium bicarbonate), and table salt (sodium chloride). You will learn how to determine how much vitamin C is in fruit juices and how much energy is associated with high-fat foods. You will also make aspirin and extract DNA from strawberries.

What you get out of your laboratory experience will depend on how much time and effort you put into it. In the real world, chemists who work in labs are not interested in simply following a procedure to get to the end of an experiment. They are interested in observing all the unexpected little things that happen along the way and in tying them all together to make sense of the results. Each and all of the experiments in this manual will give you the opportunity to apply the chemistry concepts you have learned to real-world applications, performing scientific procedures that allow you to observe reactions and phenomena for yourself and to draw your own conclusions.

Once again, welcome to UniChem International! We hope that you have as much fun performing these experiments as we did writing them.

Using This Laboratory Manual

Each experiment starts with a **Pre-Lab Assignment** to be completed before coming to lab. This assignment may require you to read the lab handout and related material in the main text, look up the physical properties of a chemical you will be using in the lab, answer some questions, or perform a calculation that will be used to analyze the data you will collect in the experiment. The answers to all Pre-Lab Assignments are recorded on your Report Sheets found at the end of each experiment. Before coming to the lab, you should also read through the list of chemicals, equipment, safety tips, and the procedure.

After the Pre-Lab Assignment, you will be given an **Introduction** to the assignment you will be doing in the lab. This could include the extraction of DNA, synthesizing polymers or aspirin, or analyzing data that were collected at a crime scene. Before you start, your supervisor will inform you of your assignment and any other special instructions you will need.

You should follow the procedure exactly how it is written, paying close attention to the weights and volumes of all chemicals needed for the experiment. Any procedural changes will be made by your supervisor during the pre-lab discussion. If you are unsure of what the procedure is asking you to do, please seek clarification before proceeding. Not following the procedure correctly may result in obtaining unusable data that are of little use to you and your partner.

Any equipment that you will need for an experiment that is not a part of the common laboratory glassware or equipment will be pointed out to you during the pre-lab discussion. Be sure to clean all equipment and return it to its original location when you are finished with it. Never put borrowed equipment in your drawer that does not belong there.

At the end of every experiment, you will find **Report Sheets** to record your data and to make a note of any observations. All data, such as weights, volumes, etc. should be recorded directly on your Report Sheet in ink. A conscientious scientist will never record important data on a paper towel or scrap paper that may be mistaken for trash and thrown away!

The end of the data section on your Report Sheet gives you an opportunity to explain any differences between what you expected to observe with your data and what you actually observed. Sometime scientists do not always get the results of the experiment they were looking for because of errors that may have been committed during the procedure. The discussion of sources of error will allow you to explain what may have gone wrong using good chemical intuition and stating how this error may have affected your results.

Next, you will reach the **Conclusion** section of the lab assignment. This may be where you report back to the person or company that hired UniChem International to perform the specific task you have just completed in the lab.

Finally, the Conclusion section is followed by **Post-Lab Questions and Problems**. These questions and problems were designed to test your understanding of the concepts from the experiment.

We hope that you will enjoy your laboratory experience and hope that this exposure to laboratory concepts will help you to understand how the concepts you are learning in the classroom are applied in the real world. We further hope that the experiments will introduce you to the exciting world of chemistry and will allow you to see the ways in which chemistry is a part of your everyday life.

And who knows—after a semester in the lab, you may decide to change careers!

Check-in and Safety

Check-in

For the majority of you, this is your first time in a chemistry laboratory. Take a look around you. Notice the fume hoods, safety equipment, balances, and other instruments and equipment you will be using this semester. The first thing we will do is to assign you to a workstation and a laboratory drawer. The lab drawer will contain most of the common glassware and equipment you will be using during this lab period throughout the semester. When you are finished with the glassware and equipment assigned to your drawer, clean it and return it to your drawer. Below this paragraph, you will find a picture and description of most of the items in your drawer. Please take the time to read this section. During the experiments, you will become use to the items in your drawer. You can save yourself some valuable lab time if you can recognize the glassware and equipment you need for a particular experiment.

Common Laboratory Equipment

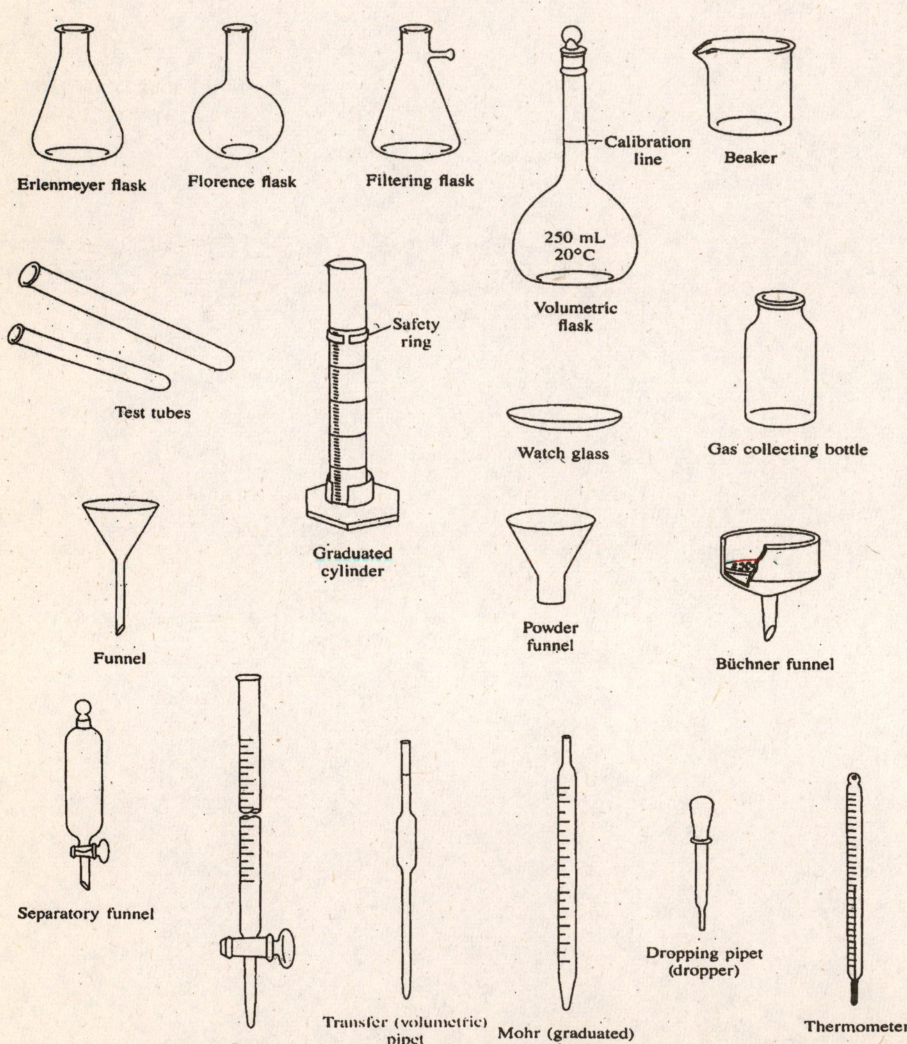

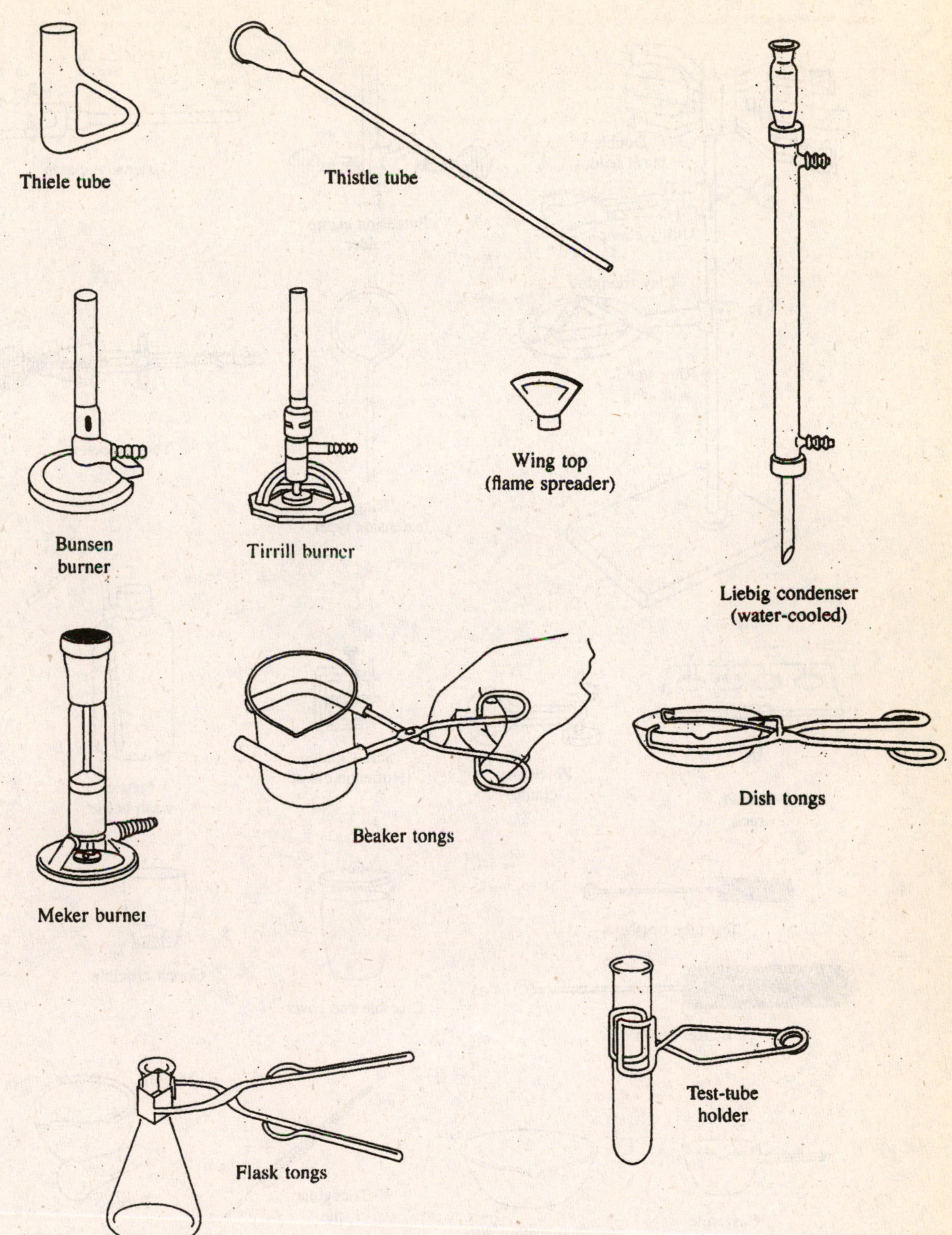
Thiele tube
Thistle tube
Liebig condenser
(water-cooled)
Bunsen
burner
Tirrill burner
Wing top
(flame spreader)
Meker burner
Beaker tongs
Dish tongs
Test-tube
holder
Flask tongs

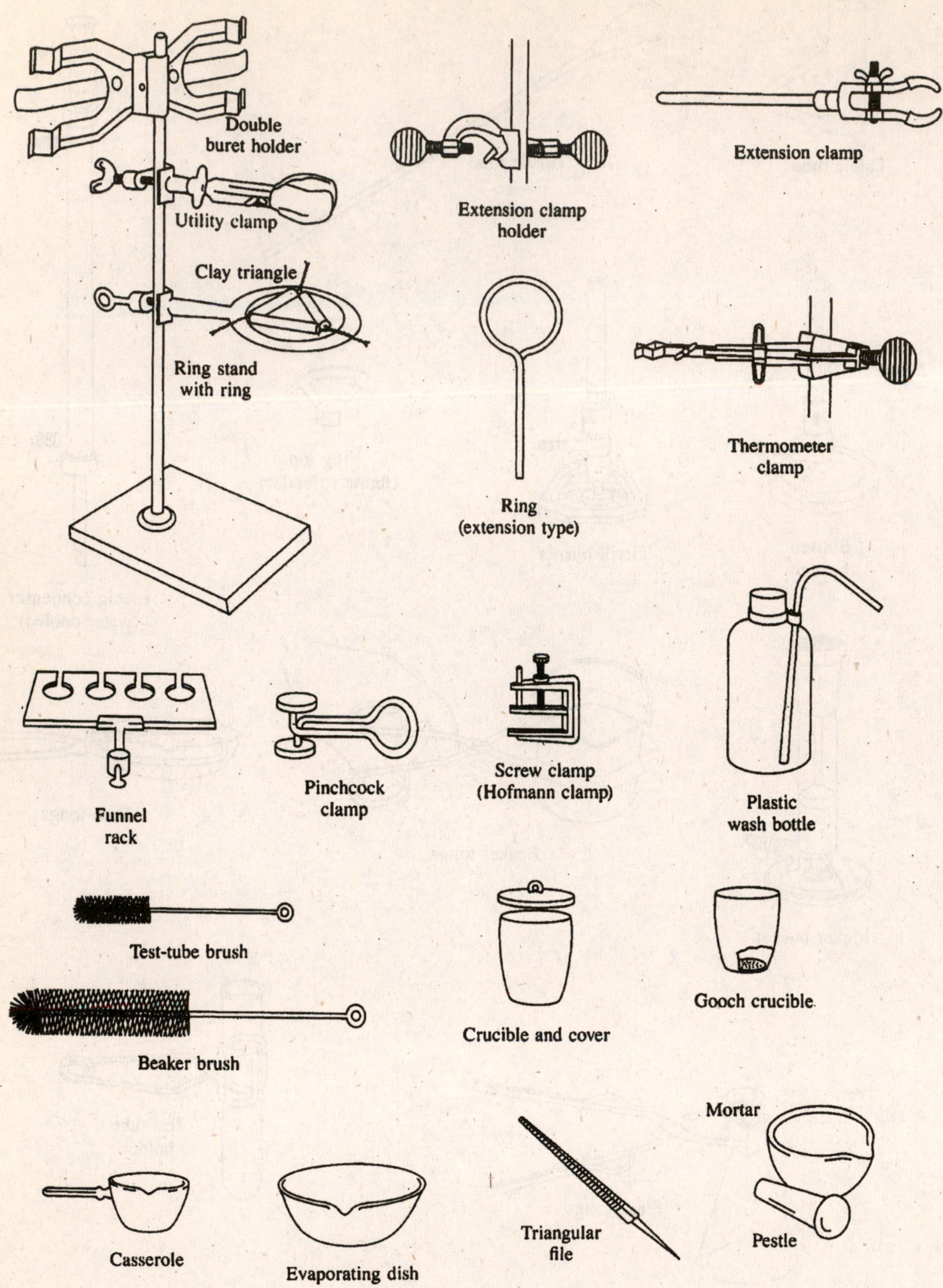
Double
buret holder
Utility clamp
Clay triangle
Ring stand
with ring
Extension clamp
holder
Ring
(extension type)
Extension clamp
Thermometer
clamp
Funnel
rack
Pinchcock
clamp
Screw clamp
(Hofmann clamp)
Plastic
wash bottle
Test-tube brush
Beaker brush
Crucible and cover
Gooch crucible
Mortar
Pestle
Casserole
Evaporating dish
Triangular
file

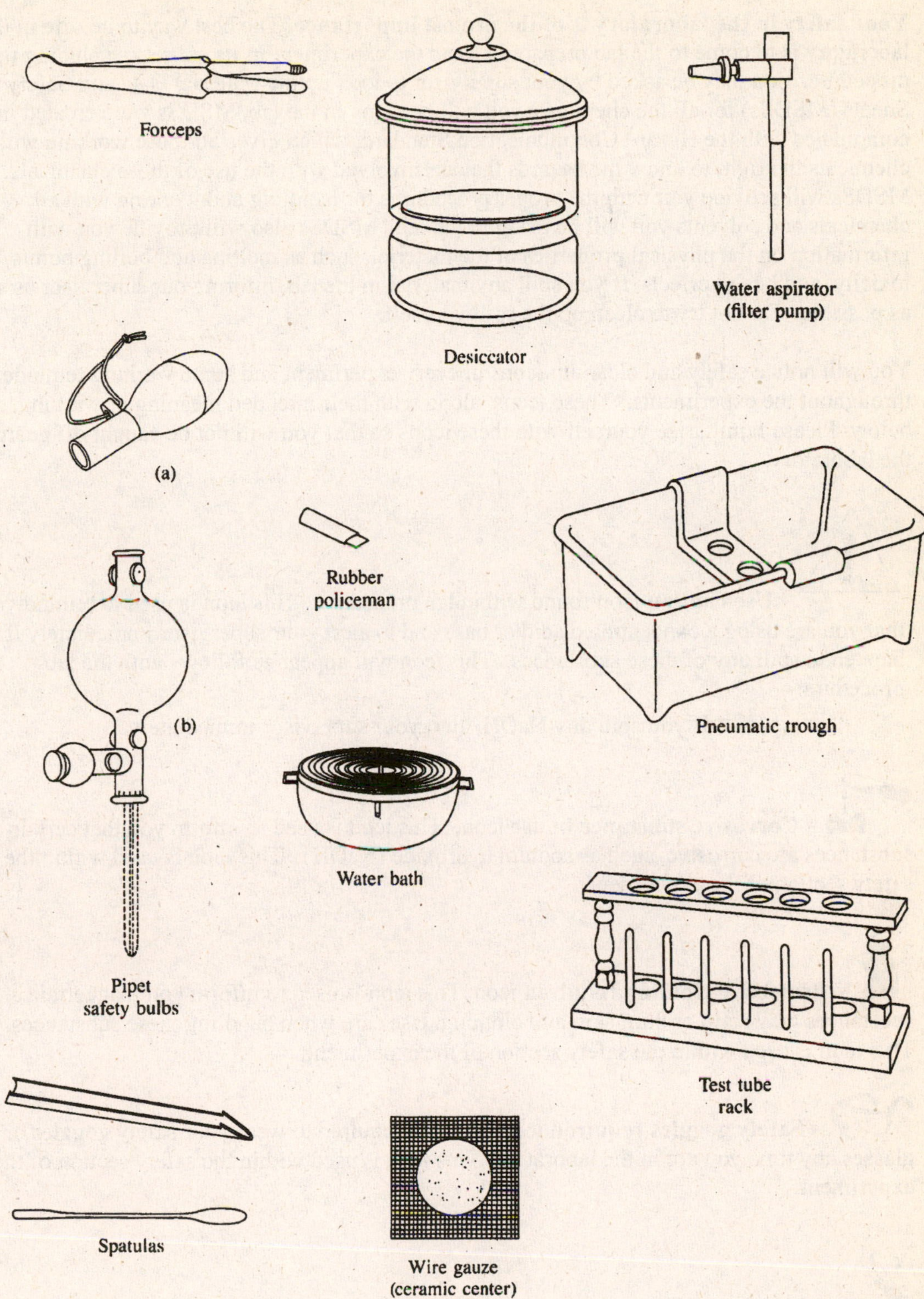
Forceps
Desiccator
Water aspirator
(filter pump)
(a)
(b)
Rubber
policeman
Pneumatic trough
Water bath
Pipet
safety bulbs
Test tube
rack
Spatulas
Wire gauze
(ceramic center)

Safety

Your safety in the laboratory is of the utmost importance. The best way to be safe in the laboratory is to come to the lab prepared. Read the experiment in its entirety, including the procedure. You may be asked by your supervisor to look up the Material Data and Safety Sheets (MSDSs) for all the chemicals you will be using in the lab. MSDSs were created in compliance with the Hazard Communication Standard, which gives all those working with chemicals the right to know the hazards that are involved with the use of those chemicals. MSDSs will provide you with the proper procedures for handling and working with all chemicals and solvents you will be using in the lab. MSDSs also will provide you with information on the physical properties of the material, such as melting and boiling points, toxicity, and health effects. If you spill any material in the lab, inform your supervisor as soon as possible. Do not try to clean up the spill yourself.

You will notice safety and clean-up icons in every experiment and some warning reminders throughout the experiments. These icons, along with their intended meaning or warning, are below. Please familiarize yourself with these icons so that you will not be caught off guard in the laboratory.

= **Use caution** icon found within lab procedure. This icon is used to remind you that you are using a concentrated acid or base and to alert your supervisor immediately if you happen to spill any of these substances. This icon will appear as follows with the lab procedure:

⚠ If you spill any NaOH, alert your supervisor immediately.

= **Corrosive substance in use** icon. This icon is used to inform you that certain substances are corrosive, such as sodium hydroxide (NaOH). This icon is used within the safety section of the experiment.

= **Protective clothing required** icon. This icon is used to inform you that certain substances may stain or burn skin and clothing. Use care when handling these substances. This icon is used within the safety section of the experiment.

= **Safety goggles required** icon. You are required to wear your safety goggles or glasses any time you are in the laboratory. This icon is used within the safety section of the experiment.

= **Hazardous fumes or strong odors produced** icon. This icon is used to inform you that certain substances can produce strong fumes or odors that may irritate the eyes, throat, and skin. You should use these substances in a fume hood or in a well-ventilated room. This icon is used within the safety section of the experiment.

= **Flame or heat in use** icon. This icon is to inform you that the experiment will require the use of an open flame. Do not have any flammable substances near the open flame. This icon is used within the safety section of the experiment.

= **Gloves required or advised** icon. This icon is to advise you to wear gloves for this experiment. Assorted sized gloves are available in the laboratory. If you do not see your size, please ask your supervisor. This icon is used within the safety section of the experiment.

= **Use tongs when handling hot glassware** icon. This icon is to advise you that there is no way to differentiate between hot glassware and cold glassware. You should never handle hot glassware with your hands, even if you are wearing disposable gloves. Test tubes holders or tongs should be used when ever glassware is heated. Remember to keep the open end of all glassware pointed away from you and those around you. This icon is used within the safety section of the experiment.

= **Dispose of used materials properly** icon. This icon is to advise you how to dispose of any left-over reagent or how to dispose of the products of the experiments. Please read and follow this section carefully. If you are unsure how to dispose of any chemical, please ask your supervisor. This icon is used within the clean-up section of the experiment.

= **Wash hands after doing this lab** icon. This icon is to remind you to wash your hands thoroughly with soap and water before you leave the laboratory. This icon is used within the clean-up section of the experiment.

In order to make the laboratory a safe working environment for all parties, the following rules and regulations must be followed at all times. Please read the following material very carefully, complete the safety quiz, and, sign the safety agreement and return it to your supervisor.

Safety Rules and Regulations

1. Always come to the lab prepared. This includes reading the experiment, and completing the pre-lab assignments.

2. Do not work alone in the laboratory. All experiments must be completed during you assigned lab period.

3. Be aware of all the safety equipment in the lab, such as emergency showers, eyewash fountains, fire blanket, fire extinguishers, fire exits, and emergency phone. Learn how the safety equipment works and the best route in case of an emergency.

4. Safety glasses must be worn whenever you are in the laboratory. Safety glasses cannot protect your eyes if they are not worn properly. You must keep your safety glasses on even if you are finished with the experiment.

5. Wear proper clothes and shoes. No sandals or open-toe shoes are allowed in the laboratory. Long sleeves should be rolled up and long hair should tied back before starting the experiment. Your supervisor will instruct you as to what is acceptable clothing in the laboratory.

6. Store all personal belongings such as book bags and coats in the space provided.

7. No food or drink is allowed in the lab at any time.

8. Use pipettes bulbs or pumps for pipetting solutions. Never pipette with your mouth or use your mouth to blow liquid out of a pipette.

9. Check the labels on the chemicals very carefully before removing any chemical from its container. Only take as much of the chemical as you need. If you take too much, never return the chemical to the original container. This may contaminate the entire bottle of the compound. Ask your supervisor how to dispose of any excess chemical or solvent.

10. Always return the caps to the reagent bottles when you are finished and return all chemicals to their proper location. If you need to lay the cap on the bench, be sure that the cap is upside down to prevent contamination. Never leave reagents at the balances or take them to your workstation.

11. Dispose of all chemicals and solvents as instructed in the clean-up section of the lab write-up or as instructed by your supervisor. Never pour any chemical down the drain unless instructed to do so. Be sure to dispose of waste in the proper containers.

12. Do not taste any of the chemicals in the laboratory, even if they are familiar to you. Never drink from any of the glassware or equipment in your lab drawer.

13. When required to note the aroma of a compound, do not smell it directly from the container. Fan your hand over the container containing the compound and then smell the vapors. Your supervisor will demonstrate the proper way to perform this technique when needed in the laboratory.

14. Most solvents used in the laboratory are extremely flammable. Never heat a flammable solvent with an open flame or have an open flame in use when flammable solvents are in use.

15. In the case of a fire involving your clothing, **STOP, DROP, and ROLL. Stop** what you are doing. **Drop** to the floor. And **Roll** around to extinguish the flames. Do not run to the fire blanket or safety shower if your clothes are on fire, **STOP DROP** and **ROLL** first. Someone else will get a fire blanket to further assist you.

16. Report any and all spills or broken glass to your supervisor immediately. Do not attempt to clean up spills by yourself. Dispose of all broken glassware in the proper

containers. Be sure to use a broom and dust pan when cleaning up broken glass. Do not use your hands. Never throw broken glass into the paper trash can.

17. To insert glass tubing into stoppers or rubber tubing, use water or glycerin as a lubricant. Wrap a towel around the glass to protect your hands. Hold the stopper between your thumb and forefinger; do not hold the stopper in the palm of your hand. Grasp the glass tube close to the end that is to fit into the stopper, and rotate the tube while pushing gently with an even pressure. Your supervisor will demonstrate the proper way to perform this technique when needed in the laboratory.

18. Stock bottles of chemicals also contain safety information. This includes chemical name, manufacturer, health, flammability, and reactivity hazards. The label also includes specific hazards unique to that chemical (e.g., oxidizer, water reactive, etc.). The diamond on some bottles indicates these hazards. The numerals in the boxes of the diamond indicate the severity of the hazard with "0" indicating little or no hazard and "4" indicating severe hazard. For example, acetone (the major component in nail polish remover) has the following ratings:

19. The **health** rating of 1 means that acetone can cause some irritation but only minor residual injury. The **fire** rating of 3 means that acetone is flammable. The **reactivity** rating of 0 indicates that acetone is stable under a variety of conditions, including exposure to water.

20. Use common sense and think! Horseplay, practical jokes, unnecessary noise, loud music, headphones, and cell phones are not acceptable. For your protection, we have to enforce all safety rules.

21. No matter how minor, report all injuries to your supervisor immediately.

22. If the fire alarm goes off while you are in the lab, calmly turn off all equipment and follow the evacuation route to exit the building. Go directly to the assigned meeting place and remain there until it is safe to re-enter the building.

23. Always clean your workstation and all common work areas before you leave the lab. Common work areas may include the fume hoods and balance areas.

24. Wash your hands before you leave the lab.

Safety Self-Quiz

The following safety quiz will test your knowledge and understanding of laboratory rules and regulations. Circle the best answer for each of the following questions. There is only one correct answer for each question. The answers can be found on page xx.

1. Safety glasses must be worn ______________.
   a. only when you feel like it
   b. only when told by your supervisor
   c. any time you are in the laboratory
   d. when chemicals are out

2. Eating is allowed in the lab ______________.
   a. at lunch time only
   b. when no one is looking
   c. when there is enough for everyone
   d. never

3. If you remove too much of a chemical you should ___________.
   a. pour it back into the original bottle or container
   b. ask your supervisor how to dispose of the excess
   c. pour it down the drain
   d. put it in your lab drawer

4. When the procedure instructs you to note the odor of a chemical, you should ________.
   a. fan the vapors toward your nose and breathe the air
   b. put the container directly under your nose and take a big whiff
   c. pour the chemical on the benchtop and then smell it
   d. put the container under your partner's nose

5. Sandals are allowed in the lab ______________.
   a. only when the outside temperature is above 80ºF
   b. if worn with socks
   c. when a lab coat is worn
   d. never because they are dangerous and never should be worn

6. When is it okay to taste chemicals in the laboratory?
   a. When you know what the chemical is.
   b. When the chemical is not hazardous.
   c. The tasting of chemicals is never allowed.
   d. When it is a brand-new bottle.

7. When is it permitted to work in the lab by yourself?
   a. You should never work in the lab alone.
   b. When you are over 18 years of age.
   c. When you have read the procedure and the MSDSs.
   d. When you are a senior and have signed the safety agreement.

8. If you take a reagent bottle to your work area, you should___________.
   a. keep it in case you need more
   b. return it as soon as you are finished
   c. put it on an adjacent bench
   d. hide it so that no one can find it

9. Open flames are allowed in the lab when _____________.
   a. flammable solvents are in use
   b. when flammable solvents are not in use near you
   c. used in the fume hood
   d. used in the balance room

10. If you spill a chemical, you should ___________.
   a. try to hide it and clean it up yourself
   b. ignore it because someone else will clean it up
   c. tell your supervisor immediately
   d. wait until you are finished with your experiment to clean it up

11. You should wash your hands ___________.
   a. only if you spilled chemicals on them
   b. only if you did not wear gloves
   c. before you leave the lab
   d. only if they are dirty

12. Hazardous waste should be ______________.
   a. poured down the drain
   b. placed in the properly labeled waste container
   c. thrown in the trashcan
   d. left on your bench

13. All personal items should be __________.
   a. left right by the exits
   b. left in the space provided
   c. placed on your benchtop
   d. left in the balance room

14. Prior to coming to lab, you should _____________.
   a. skip the lab and just answer the post-lab questions
   b. call your lab partner and ask what the lab is about
   c. read the entire experiment and compete the pre-lab assignment
   d. figure it out once you get into lab

15. If your hair or clothing catches on fire, you should __________.
   a. run to the safety shower to put out the flames
   b. Stop, Drop, and Roll
   c. run to get the fire blanket
   d. pull the fire alarm

16. If you cut your hand you should ________________.
    a. wrap the cut with a paper towel
    b. go to the bathroom to clean it
    c. inform your supervisor immediately
    d. don't worry about it because it has stopped bleeding

17. When is it okay to "goof off" in the laboratory?
    a. When you are finished with your experiment.
    b. Before the experiment starts
    c. When your supervisor is not looking.
    d. It is never okay to goof off in the lab because an injury or an accident may occur.

18. If the fire alarm goes off while you are in lab, you should ______________.
    a. exit the building calmly and quickly and go directly to the designated meeting place
    b. go get lunch
    c. go to the library
    d. go home

19. On the first day of lab, you should ______________.
    a. show up late because you never have lab on the first day
    b. familiarize yourself with the use and location of all the safety equipment in the laboratory
    c. not listen because you have had other labs before
    d. think about what you are going to eat for dinner

20. The best way to be safe in the laboratory is to ________.
    a. come to the lab prepared
    b. read the experiment once you get to the lab
    c. don't go to the lab
    d. let your lab partner do all the work

Answers to Safety Quiz

1. c	2. d	3. b	4. a	5. d	6. c	7. a	8. b	9. c	10. c
11. c	12. b	13. b	14. c	15. b	16. c	17. d	18. a	19. b	20. a

Safety Agreement

I, ______________________________, agree to follow all the safety rules for conduct, handling of chemicals, and use of equipment that have been covered in the "Laboratory Safety" handout as well as any others that are prescribed by the supervisor throughout the course of the year. I have read the safety rules and had them explained and demonstrated to me by my supervisor and have been instructed in the use of safety equipment in the classroom. I understand that failure to follow the above-described rules and procedures may result in my exclusion from further participation in lab work and may result in my dismissal from this course.

Student signature: ___________________________________

Date: ______________________

Course title: ________________________

Section: ___________________________

Supervisor: _______________________

Please return a signed copy of this safety agreement to your supervisor. Signing this agreement implies that you will follow all laboratory rules and regulations.

Experiment 1

Measurement of a Missing Object

PRE-LAB ASSIGNMENT

Reading:

1. Chapter 1 and Appendix of *Adventures in Chemistry*.
2. Preface and safety rules and requirements from Lab Manual.
3. This experiment.

INTRODUCTION

Welcome to your first day at UniChem International, a major chemical conglomerate with departments in forensics, chemical analyses, criminal division, synthesis and biochemistry laboratories, etc. Your first assignment as a member of the Criminal Forensic Lab (CFL) is to investigate several items that have been reported missing from the local art museum. Our special forensic division has been requested to determine the mass of a missing art object, one of a cubic trio usually located at the front entrance of the local art museum.

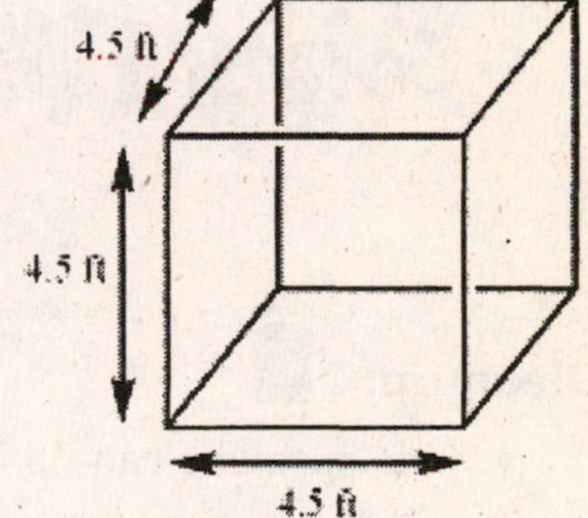

Figure 1.1

Working in teams of two, investigators will follow the given **quantitative** method to determine the measurement of the missing cube of pure iron. The curator's records indicate that each cube was identical, each measuring 4.5 feet on each side and made of pure iron. A sketch of the missing cube is shown in Figure 1.1. CFL has been asked to help approximate the mass of the stolen cube in order to help police estimate the possible number of people involved in the theft and the size of vehicle needed to transport the cube from the crime scene. This will be accomplished using scientific measurement, including measuring mass with a balance, determining the volume by water displacement and accounting for errors in the measurements. In this way, we can approximate the mass of the cube. Density is defined as mass per unit volume, where the mass is usually expressed in grams (g) and the volume in milliliters (mL). This can be represented mathematically by the equation:

$$\text{Density} = \frac{\text{mass of object (g)}}{\text{volume of object (mL)}}$$

Your first assignment involves determining the density of smaller pieces of iron in the laboratory, this will be used to determine the mass of the missing art object.

Quantitative Analysis In the strict chemical sense, this has to do with designing a chemical test to allow the measurement of amounts and proportions of the components of any given substance or mixture. In a broad context, this type of analysis typically involves devices for measuring length, mass, volume, pressure, temperature, pH, etc.

LEARNING OBJECTIVES

Be able to:

- Perform quantitative analysis in order to determine the density of iron.
- Analyze data using a graphing program such as Excel®.
- Perform error analysis.

APPARATUS

Chemicals: None needed for this experiment.

Equipment:

- Five different and assorted pieces of iron
- Assorted sizes of graduated cylinders (actual sizes will depend on the iron pieces)
- Analytical balances
- Computer with Excel® (or other graphing program)

- Are your safety glasses on?
- Do not drop iron pieces into the gradated cylinders because this may cause the graduated cylinder to shatter.
- To avoid this, tilt the graduated cylinder to one side and allow the metal piece to gently slide down the side of the cylinder.

Clean-up: 

- No special clean-up is necessary for this experiment. All liquid can go down the drain.
- Iron pieces should be placed on a towel until dry, after which they may be stored until needed again.

PROCEDURE 1

Determining the Density of Iron

1. Obtain at least five different samples of iron from those available in the lab.
2. Suggested samples of iron are five nails, one large iron bolt, one medium sized of iron rod, etc.
3. Weigh each different sample of iron on the analytical balance to determine its mass to the nearest 0.001 g and record the mass for each type of iron on your report sheet.

- Obtain a graduated cylinder to determine the volume of the various types of iron by water displacement.
- The graduated cylinder should be large enough to cover the various types of iron with water.

4. Fill the graduated cylinder halfway with water.
5. Record the initial volume of water (before the iron is added) to the nearest 0.1 mL on your report sheet.
   - Always record the value of the graduated cylinder at the bottom of the meniscus. (See Figure 1.2)
6. Carefully tilt the graduated cylinder sideways so that no water spills out, and gently slide one piece of iron into the cylinder.
7. Be sure that the entire piece of iron is covered with water and dislodge any air bubbles by gently tapping the sides of the cylinder.
8. Record the final volume of water (after iron is added) on your report sheet. (See Figure 1.2)
9. Carefully pour the water out of the graduated cylinder and set the piece of iron aside to dry.
10. Repeat steps 7 – 13 with the remaining four samples of iron.
11. Use Excel® to help calculate the change in volume, density for each piece of iron, average density, standard deviation, experimental error, and percent relative uncertainty (See *Error Analysis in the Chemistry Laboratory* in the Appendix at the end of the Lab Manual).

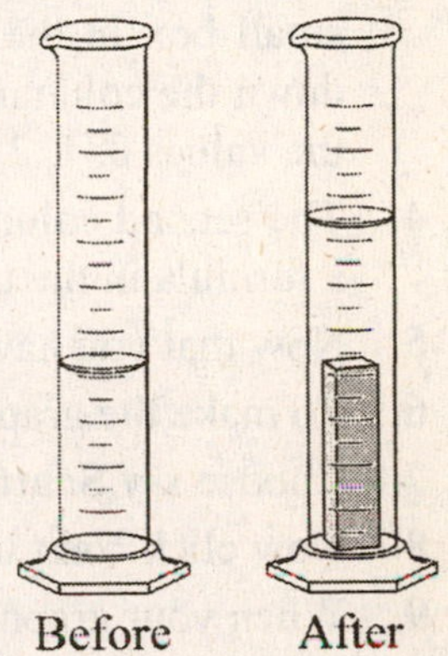

Figure 1.2 Meniscus before and displacement

Refer to Procedure 1 Data Table of your Report Sheet on p. 5.

PROCEDURE 2

Using Excel®

1. You are now ready to use Excel® to do some timesaving calculations.
2. Record the sample of iron measured in the first column.
3. In the second column, enter the mass of the object (you can denote grams with **g**). Only enter numeric values, not units, in the cells below your header.
4. In the third and fourth columns, enter your volume before and after displacement respectively (you can denote milliliters with **mL**).
5. The fifth column will be used to calculate the change in volume. To do this you will enter a formula in the top cell of this column and then copy the formula down the column where applicable. (Ask for help from your supervisor, if needed.)
6. The sixth column will be used to calculate the density of the piece of iron.
7. The seventh column will be used to calculate the average density of iron.
8. In the eighth column, calculate the standard deviation of your measurements. The standard deviation indicates your analytical precision.
9. Knowing the density and the volume of iron, find the mass of the stolen iron cube.

Refer to Procedure 2 Graphing Data of your Report Sheet on p. 5.

PROCEDURE 3

Graphing

1. Now make a graph of the number of cubes versus the mass of the cubes in Excel®.
2. Create a graph that correlates the number of iron cubes (make a graph for up to six cubes) versus the total mass.
3. When graphing, Excel® reads the first column as the x variable; therefore enter the number of iron cubes in column 1. Here is a handy trick when you want to enter a series of values down a column such as 0, 1, 2, etc. You only need to enter the value 0 in the first cell and then the value 1 in the cell below. Highlight these two cells, click on the small box in the lower right corner of your highlighted area, and then drag the mouse down the column. This automatically fills a column with the correct values (you now see the values 0, 1, 2, 3, etc.).
4. The second column will be the y variable, which is the mass of iron; again, you can type a formula in the uppermost cell and copy it down through the column.
5. Now that you have created the table, you are ready to make the graph using Excel®.
6. To make the graph, click on the graphing icon in the uppermost tool bar.
7. Choose **x-y Scatter** as the graph type.
8. Now click **Next** to be able to label the axes (don't forget the units) and title.
9. When your graph is completed, click on **Finish**.
10. Now click on **Chart** and select **Add a Trend Line.** Click the **Options** tab, and then click on the box next to **Display Equation on Chart**. You also will want to check off the box next to **Display R-squared Value on Chart** and print the graph.
11. Use your graph to report to the police the weight capacity required to haul off any given number of iron cubes. The R^2 value for your graph is called the **correlation coefficient**, and indicates how well the data points fit the type of curve you chose. The closer this value is to 1.000, the better your data points fit a linear correlation.

Attach Graph to your Report Sheet on p. 8.

NAME (print) ______________________________________ DATE ______________
LAST FIRST

LABORATORY SECTION ______ PARTNER(S) ______________________________________

REPORT SHEET

PROCEDURE 1

Data Table 1

Sample of Iron	Mass (g)	Initial Volume (mL)	Final Volume (mL)

PROCEDURE 2

Graphing Data Section

Fill in the blanks with the data obtained from Excel®.

Average density ______________ g/mL

Standard deviation ______________

Experimental error ______________ %

Percent relative uncertainty ______________ %

Mass of stolen cube ______________ g

Size of Vehicle ______________

NAME (print) ______________________________ DATE ______________
LAST FIRST

LABORATORY SECTION ______ PARTNER(S) ______________________________

DATA ANALYSIS

Account for your accuracy by finding the literature value for the density of iron (chemical symbol Fe) in a chemistry reference book or on the Internet (be sure to cite the complete reference for where you found this information). Compare this value to your experimentally determined value, and calculate your experimental error below. Experimental error = literature value – experimental value/literature value * 100%. What is your percent relative uncertainty of the experimental density you determined for iron? Percent relative uncertainty = standard deviation/mean * 100%.

SOURCES OF ERROR

Once the data have been collected, the possibility of error in the results must be addressed. Please cite at least two random sources of error and briefly explain how each source of error may have affected your data. For a review of error analysis, see *Error Analysis in the Chemistry Laboratory* in the Appendix at the end of this manual.

NAME (print) ____ DATE ____________________
LAST FIRST

LABORATORY SECTION ______ PARTNER(S) ____

POST-LAB QUESTIONS

1. How many significant figures should you put on the estimated mass of the iron cube value you are reporting to the police? Briefly explain.

2. Using the equation of the line from your graph, calculate the number of cubes that a 300-ton capacity rental truck can carry.

3. The artist of the cubic trio is considering building another set of cubes, but to reduce the cost of shipping, aluminum will be used. Determine the mass of a pure aluminum cube of the same dimensions.

4. Which will displace water in a graduated cylinder more, an aluminum bolt or an iron bolt (assume that both bolts are identical in size)? Explain your answer.

5. The recommended daily allowance of iron is approximately 18 mg. Wheaties® and Cheerios® both contain approximately 45% of the daily recommended allowance of iron in a 1-cup serving. How many milligrams of iron are in a serving? How many grams of iron are in a serving?

NAME (print) ______________________________ DATE ______________
LAST FIRST

LABORATORY SECTION ______ PARTNER(S) ______________________________

CONCLUSION

Based on your analysis how close was your value for the density of iron to the literature value? How much did the missing cube weigh and what size vehicle was needed to remove the cube? What other things can you conclude or state? State your recommendations and the reasoning behind them in complete sentences. Be sure that your conclusion is complete and concise.

Attach Graph here:

Experiment 2

Analysis of Star Shells and Fireworks

PRE-LAB ASSIGNMENT

Reading:

1. Read the experiment.

Assignment:

2. Obtain an empty cereal box (or similar), and bring it to lab.
3. Review how electronic energy level transitions produce emission spectra.
4. Obtain a chart of the electromagnetic spectrum for reference

Questions:

5. Answer the Pre-Lab Questions.

INTRODUCTION

Fort Harrison was established as an Army base during the Civil War and served as a training ground for infantry troops and a munitions repository until it was decommissioned following the end of the Cold War in the 1990s. After nearly a decade of debate, community leaders in Harrison County finally have decided to turn the base into a historical site and nature preserve.

Before the Fort Harrison conversion can begin, however, the unsalvageable buildings need to be demolished, and all leftover ordnance must be disposed of properly. A large quantity of old explosives, including star shells (which had been used to illuminate battlefields at night, as well as to pass signals) and other fireworks, has been found in several warehouses on the base. Along with highly reactive compounds such as potassium chlorate ($KClO_3$), the star shells and fireworks also contain large amounts of metal salts, which pose potential health and environmental risks.

The Harrison County Supervisors have hired UniChem International to analyze and treat the munitions found in Fort Harrison. The Analytical Chemistry Division of UniChem International will identify the metal-containing compounds and assess the condition of the reactive substances. Once this analysis is complete, UniChem's Environmental Remediation specialists then will be able to safely remove and properly dispose of the compounds contained in these munitions.

As one of UniChem's top analytical chemists, you have been given the task of determining the identities of the unknown metal-containing salts from comparison of their spectra with the spectra of known metal-containing compounds. In addition, you also may be asked to perform a small-scale test to determine if the $KClO_3$ found in the old munitions is still reactive.

LEARNING OBJECTIVES

Be able to:

- Construct a spectroscope and use it to obtain visible emission spectra.
- Analyze visible spectra to determine the identity of an unknown substance.
- Explain how spectra are produced, why they consist of discrete lines, and how they can be used to identify specific substances.

APPARATUS

Chemicals:

- Metal-containing salts (e.g. LiCl, KCl, NaCl, $BaCl_2$, $CuCl_2$, $SrCl_2$, etc.)
- Unknown metal-containing salt samples from the star shells/fireworks
- 6.0 M hydrochloric acid (HCl)
- Potassium chlorate ($KClO_3$) samples from the star shells/fireworks
- Sugar
- Concentrated (18 M) sulfuric acid (H_2SO_4)

Equipment:

- Empty cereal box or similar cardboard box
- Small piece (about 2 × 2 cm) of diffraction grating material
- Black electrical tape
- Scissors and/or safety razor blade
- Paper ruler
- Nichrome wire on a handle
- Porcelain spot plate

- Are your safety glasses on?
- Do not cut yourself with the scissors or safety razor.
- Do not burn yourself with the nichrome wire or metal salts.
- Clean the nichrome wire with HCl in the fume hood.
- HCl and H_2SO_4 can burn your skin and clothes.
- Procedure 4 must be done in the fume hood with the door down. The reaction may spatter very hot materials and may break the test tube.

Clean-up: 

- Please dispose of all waste as directed by your supervisor.
- Wash your hands thoroughly with soap and water before you leave the laboratory.

PROCEDURE 1

Constructing a High-Tech Spectroscope

1. Carefully cut an opening (about 2 cm high and 1 cm wide) close to the edge of one end of the box (at *A*). Attach two pieces of electrical tape vertically to create a 1-mm-wide slit at *A*. Refer to Figure 2.1.
2. Carefully cut another opening (about 1 × 1 cm) close to the edge of the other end of the box and directly opposite *A* (at *B*).
3. Obtain a piece of diffraction grating (about 2 × 2 cm). Hold it by the edges only, and be careful not to get any fingerprints on the diffraction grating surfaces.

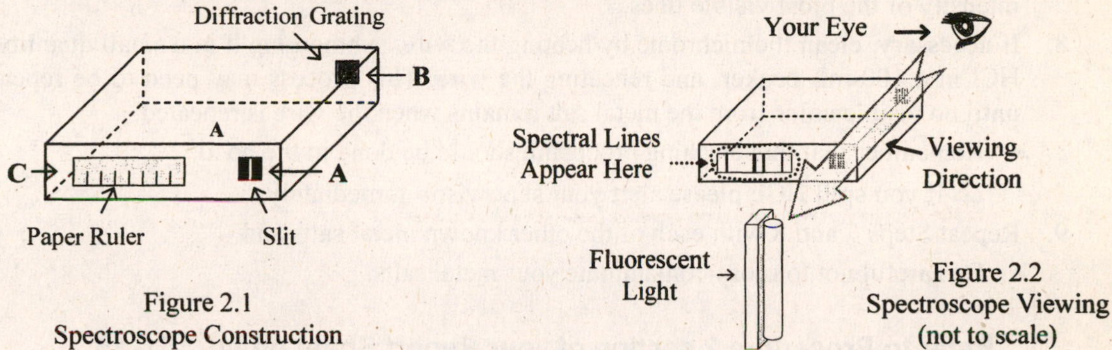

Figure 2.1
Spectroscope Construction

Figure 2.2
Spectroscope Viewing
(not to scale)

4. Carefully place the diffraction grating over the 1-cm-square hole (*B*) with the sides of the diffraction grating and the sides of the hole parallel. Point the box at a fluorescent light and look through the diffraction grating and the slit (*A*). If you line up the light, the slit, and the diffraction grating, you will see a spectrum either off to the side or above or below the slit.
5. If the spectrum is to the right or left of the slit, then the diffraction grating is properly oriented. If the spectrum is above or below the slit, rotate the grating 90 degrees to achieve the correct orientation. Adjust the diffraction grating until the spectral lines are perpendicular to the front edge of the box (and as parallel as possible to the slit). Use electrical tape to carefully tape the grating in place. Refer to Figure 2.2.
6. Again look at a fluorescent light through your spectroscope. Carefully cut a rectangular hole (about 2 × 15 cm) such that the complete spectrum falls "within" the hole, with several centimeters of open space on either side (at *C*).
7. Tape a paper rule over *C* so that the scale readings are visible when observing a spectrum.
8. Tape closed any openings that would allow light to enter the spectroscope (other than *A*, *B* and *C*, of course!).

PROCEDURE 2

Observing Known Metal-Salt Spectra

1. Obtain a porcelain spot plate. Label the wells with the formulas of the metal salts.
2. Use a scoopula to transfer small quantities of each salt to the appropriate wells. Be careful not to cross-contaminate the substances.
3. Light a Bunsen burner, and adjust it until you have a crisp blue flame with a distinct inner cone. Make sure that the top of the Bunsen burner does not have any spilled substances that will burn and affect the color of the flame.

4. Make a small loop (about 2 mm in diameter) in the end of the nichrome wire.
5. Hold the loop in the Bunsen burner flame. Look through your spectroscope at the glowing nichrome wire. Describe what you observe both with your naked eye and through the spectroscope.
   - Do not get so close that you catch this high-quality instrument on fire!
6. Remove the nichrome wire from the flame, and place the loop into one of the metal salts in the spot plate. A small clump of the compound should adhere to the hot metal.
7. One person will need to hold the clump of metal salt in the flame while another person looks through the spectroscope. Describe the appearance of the metal salt while in the flame. Sketch the emission spectrum for this metal salt, indicating the color, location and intensity of the most visible lines.
8. If necessary, clean the nichrome by heating the wire, submerging it in a small quantity of HCl in a 100-mL beaker, and reheating the wire. This process may need to be repeated until no visible color from the metal salt remains when the wire is reheated.
   - Remember that this cleaning procedure should be done in the hood.

     ⚠ If you spill HCl, please alert your supervisor immediately.
9. Repeat Steps 7 and 8 with each of the other known metal salts and
   - Be careful not to cross-contaminate your metal salts.

Refer to Procedure 2 section of your Report Sheet on pp. 14 – 15.

PROCEDURE 3

Identifying the Unknown Metal Compound from the Star Shells

1. Obtain the unknown metal-containing salts you have been assigned to test. Carefully transfer small quantities of each into clean labeled wells in the spot plate. Record the Identification Code of each substance, and describe the physical appearance of each.
2. Repeat the flame-test procedure used with the known metal-containing salts to determine the spectra of the unknown metal salts. Record both the flame color and spectrum observations for each unknown.

Refer to Procedure 3 section of your Report Sheet on pp. 15 – 16.

PROCEDURE 4

Testing the Reactivity of $KClO_3$

1. In a small beaker, combine approximately 1.5 g of sugar, 1.5 g of $KClO_3$ recovered from the star shells/fireworks, and 1 g of your favorite metal-containing salt. Use a stirring rod to mix the substances thoroughly.
2. Transfer the contents of the beaker to a medium Pyrex test tube. Inside a fume hood, clamp the test tube to the ringstand, with the test tube angled toward the back of the hood.
3. Add a small squirt of concentrated H_2SO_4 to the contents of the test tube, and quickly close the fume hood door. Record your observations.

   ⚠ If you spill any H_2SO_4, please alert your supervisor immediately.

Refer to Procedure 4 section of your Report Sheet on p. 16.

NAME (print) ______________________________________ DATE ____________
LAST FIRST

LABORATORY SECTION ______ PARTNER(S) ______________________

REPORT SHEET

PRE-LAB ASSIGNMENT

Pre-Lab Questions

1. Consider electromagnetic radiation with the wavelengths or frequencies given below.
   a. Classify each according to the type of radiation.
   b. Is this radiation visible? If it is visible, give its color.

Wavelength/Frequency	Type of Radiation	Visible? Color?
$\lambda = 8.3 \times 10^{-9}$ m		
$\lambda = 6.1 \times 10^{-7}$ m		
$\nu = 3.9 \times 10^{8}$ Hz		
$\nu = 9.5 \times 10^{15}$ Hz		

2. Briefly but completely describe what occurs to produce an emission spectrum. (Your answer should include such words as *electrons, energy, quantized, photons*, etc.)

3. Briefly but completely describe why the emission spectra of different elements act as chemical "fingerprints" that can be used to identify a given substance.

NAME (print) ______________________________ DATE ______________
LAST FIRST

LABORATORY SECTION ______ PARTNER(S) ______________________________

PROCEDURE 2

Metal-Salt Emission Spectra

1. Record your observations for the known metal-containing salts and their spectra.

Metal-containing salt _________ Metal ion in compound _______ Physical description of metal salt	Color of metal salt in flame ____________ Sketch of spectrum
Metal-containing salt _________ Metal ion in compound _______ Physical description of metal salt	Color of metal salt in flame ____________ Sketch of spectrum
Metal-containing salt _________ Metal ion in compound _______ Physical description of metal salt	Color of metal salt in flame ____________ Sketch of spectrum
Metal-containing salt _________ Metal ion in compound _______ Physical description of metal salt	Color of metal salt in flame ____________ Sketch of spectrum

NAME (print) ______________________________ DATE ____________
LAST FIRST

LABORATORY SECTION ______ PARTNER(S) ______________________________

Metal-containing salt ________ Metal ion in compound ______ Physical description of metal salt	Color of metal salt in flame ______________ Sketch of spectrum
Metal-containing salt ________ Metal ion in compound ______ Physical description of metal salt	Color of metal salt in flame ______________ Sketch of spectrum

PROCEDURE 3

Unknown Salt Emission Spectra

1. Record your observations for one unknown metal-containing salt and its spectrum.

Metal-containing salt ________ Metal ion in compound ______ Physical description of metal salt	Color of metal salt in flame ______________ Sketch of spectrum

2. Identify the metal(s) in this unknown. (The unknown may contain one or two metal ions.) Briefly explain your reasoning.

NAME (print) ______________________________ DATE ______________
LAST FIRST

LABORATORY SECTION ______ PARTNER(S) ______________________________

3. Record your observations for another unknown metal-containing salt and its spectrum.

Metal-containing salt ________ Metal ion in compound _______ Physical description of metal salt	Color of metal salt in flame ________ Sketch of spectrum

4. Identify the metal(s) in this unknown. (The unknown may contain one or two metal ions.) Briefly explain your reasoning.

PROCEDURE 4

Testing the Reactivity of $KClO_3$

1. Record your observations from the reactivity test.

2. After all the years the munitions have been stored, can the $KClO_3$ be considered decomposed, inert and therefore safe to handle, or is it still reactive? Explain.

NAME (print) ______________________________ DATE ______________
LAST FIRST

LABORATORY SECTION ______ PARTNER(S) ______________________________

POST-LAB QUESTIONS

1. Which is likely to be the better way to identify an unknown substance by its physical appearance or by its emission spectrum? Explain, citing specific examples if possible.

SOURCES OF ERROR

Once the data have been collected, the possibility of error in the results must be addressed. Please cite at least two random sources of error and briefly explain how each source of error may have affected your data. For a review of error analysis, see *Error Analysis in the Chemistry Laboratory* in the Appendix at the end of this manual.

NAME (print) ______________________________ DATE ______________
LAST FIRST

LABORATORY SECTION ______ PARTNER(S) ______________________________

CONCLUSION

Based on your analysis, what metals were contained in the unknown samples you analyzed? Be sure that your conclusion is complete and concise and written in sentence format.

Experiment 3

Electrochemical Cells

PRE-LAB ASSIGNMENT

Reading:

1. Read the experiment.

Research:

2. Review oxidation–reduction reactions and electrochemical cells (see Chapter 3 of *Adventures in Chemistry*).

Question:

3. Answer the Pre –Lab Questions.

INTRODUCTION

In an electrochemical cell, an oxidation–reduction reaction takes place and produces electrical energy. When designed and properly constructed to meet the appropriate performance requirements, an electrochemical cell can be used to start a car, run an iPod, or power a heart pacemaker. Billions of these electrochemical cells, known as **batteries**, are produced annually.

The simplest electrochemical cells are made by combining two half-cells. Each half-cell consists of a piece of metal in contact with a solution of that metal's cation (i.e. Cu metal in a solution of Cu^{2+}). When the metal pieces from two different half-cells are connected electrically, and the metal ion solutions are in contact via a porous barrier or salt bridge, a complete circuit is established. Then electrons will flow through the wire from the substance that more easily loses electrons to the substance that more easily gains electrons.

This type of simple electrochemical cell typically does not constitute a useful battery. Practical batteries must have a high enough potential difference (voltage) and produce large enough currents to perform their desired functions. To achieve this, it may be necessary to connect several electrochemical cells and to manipulate the constitution of the specific substances. In addition, real batteries need to be durable, small and light enough, long lasting, safe, and not too expensive. These requirements can be met by a combination of basic scientific research into the oxidation–reduction characteristics of different substances and the engineering aspects of combining substances appropriately to produce batteries that meet people's needs.

In order to stay competitive in the global battery market, UniChem International has decided that it needs to expand the research program in its Electrochemical Division. In this program, electrochemists will investigate a variety of substances to determine their relative oxidation–reduction abilities. These researchers also will investigate various aspects of battery design, such as how to connect electrochemical cells to achieve higher voltages and currents. Eventually, the engineers will be called on to translate this basic research into the production of useful batteries that can increase UniChem International's share of the global battery market.

As one of UniChem's most experienced electrochemists, you have been given a highly coveted position on the Electrochemical Research Team (ERT). You and your fellow ERT members have been asked to construct simple lead-acid electrochemical cells, similar to those found in car batteries, and evaluate factors affecting their performance characteristics. In addition, you will investigate a number of promising metals to determine some possible combinations that will meet the voltage requirements of particular battery designs.

LEARNING OBJECTIVES

Be able to:

- Create electrochemical cells, both voltaic and electrolytic.
- Describe the components of simple electrochemical cells and how they function.
- Analyze the effects of various factors (e.g., electrode surface area, charging time, whether cells are connected in series or parallel, etc.) on a battery's performance.
- Measure potential differences between different substances and interpret them in terms of battery design.

APPARATUS

Chemicals:

- Lead strips
- 1.0 M sulfuric acid (H_2SO_4) solution
- Small pieces of various metals (e.g. Fe, Cu, Zn, Sn, Ag, Al, and Ni)
- 1.0 M solutions of various metal ions (e.g. Fe^{2+}, Cu^{2+}, Zn^{2+}, Sn^{2+}, Ag^{+}, Al^{3+}, Ni^{2+}, and K^{+})

Equipment:

- Power supply
- Wires with alligator clips
- Filter paper
- Scissors
- Digital multimeter
- Small motor

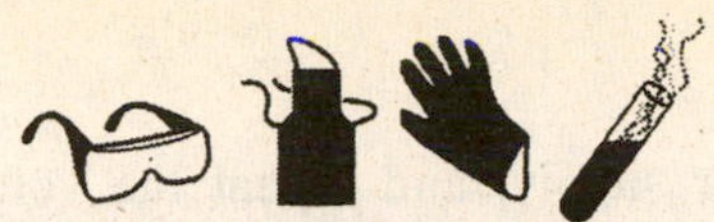

- Are your safety glasses on?
- Sulfuric acid can burn your skin and clothing.
- You may want to wear gloves to handle the lead strips.
- Charge the lead-acid cell in a well-ventilated area or in a fume hood.
- Ag^+ solution can stain your skin and clothing.

Clean-up:

- Dispose of all excess reagents and waste as directed by your supervisor.
- Wash your hands thoroughly with soap and water before you leave the laboratory.

PROCEDURE 1

Lead-Acid Electrochemical Cells

In the first part of this research project, you will create lead-acid electrochemical cells and investigate the effect of charging time on the cell's ability to provide electrical energy.

1. Obtain two lead strips (one labeled + and one labeled -), two wires with alligator-clip ends (one red and one black), and a power supply.
2. Add about 150 mL of 1.0 M H_2SO_4 to a 250-mL beaker.
3. Place the lead strips on opposite sides of the beaker, with the bottoms submerged, as shown in Figure 3.1. Bend the tops of the strips to hold them in place, as shown.

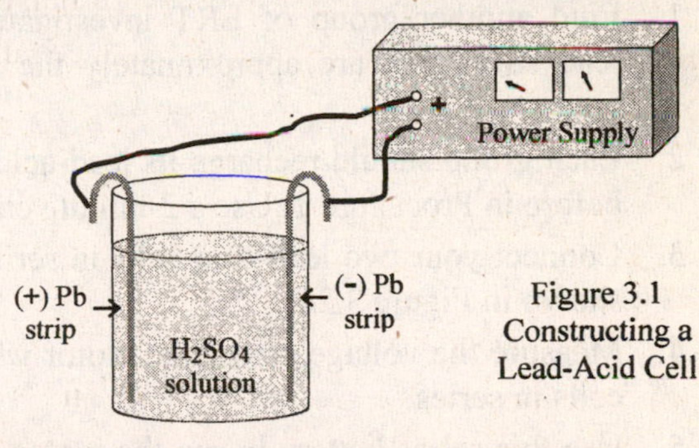

Figure 3.1 Constructing a Lead-Acid Cell

4. Use a ruler to estimate the submerged dimensions of each of the lead strips.
5. Clip one end of the red wire to the + lead strip and the other end to a positive (red) jack on the power supply. Clip one end of the black wire to the – lead strip and the other end to a negative (black) jack on the power supply.
6. Turn on the power supply, as directed, to charge the lead-acid storage cell. You should see evidence of reaction occurring at each of the lead electrodes. Charge the cell for exactly 1 minute. Record your observations.

   ⚠ Remember to work in a well-ventilated room or a fume hood. Do not breathe the vapors produced while the cell is charging.
7. Unclip the wire leads from the two electrodes. Use the multimeter, as directed, to measure the potential difference (voltage) between the two lead electrodes.
8. Connect the battery to a small motor. Measure how long the motor runs.
9. After the motor has stopped running, again use the multimeter to measure the voltage of the battery.
10. Reconnect the battery to the power supply, and recharge for exactly 2 minutes.

11. Disconnect the power supply, and repeat the voltage measurement on the recharged battery.
12. Connect the recharged battery to the motor, and measure how long the motor runs this time.
13. Once again, reconnect the battery to the power supply. This time, recharge for exactly 3 minutes.
14. Disconnect the power supply, and repeat the voltage measurement on the recharged battery.
15. Connect the recharged battery to the motor, and measure how long the motor runs this time.

Refer to Procedure 1 section of your Report Sheet on pp. 25–26.

PROCEDURE 2

Lead-Acid Storage Battery Characteristics

In this part, you will combine with other ERT researchers to investigate the effect of electrode surface area and combining cells in series or parallel on this type of battery's ability to do work.

1. Find another group of ERT investigators that used lead strips that are approximately the same size as yours.
2. Each group should recharge its lead-acid cell as done before in Procedure 1. Use a 2-minute charging time.
3. Connect your two lead-acid cells in series (+ to –) as shown in Figure 3.2a.
4. Measure the voltage across the motor when using the cells in series.
5. Use this series battery to run the motor, and measure how long it runs.
6. Recharge both cells for 2 minutes, and then connect the cells in parallel (+ to +, – to –) as shown in Figure 3.2b.
7. Measure the voltage across the motor when using the cells in parallel.
8. Use this parallel battery to run the motor, and measure how long it runs.
9. If instructed by your supervisor, combine with more groups of ERT researchers and repeat this procedure with (up to) six cells in series and in parallel.

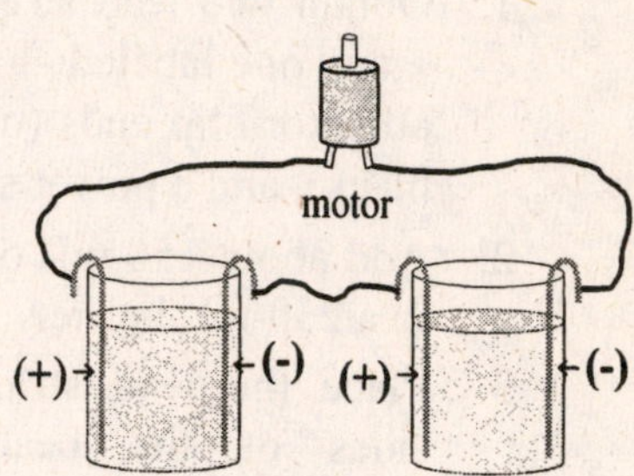

Figure 3.2a
Electrochemical cells in series

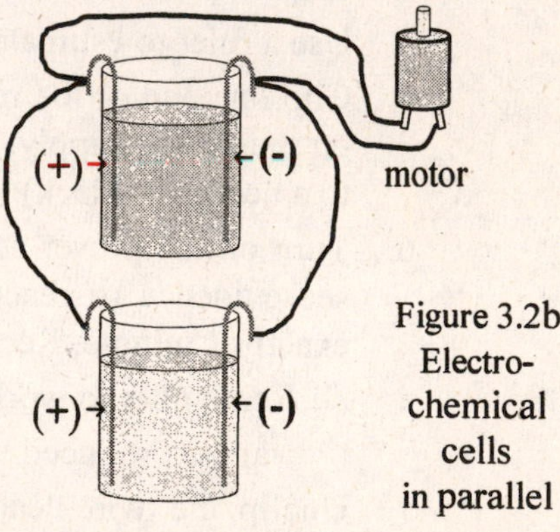

Figure 3.2b
Electrochemical cells in parallel

Refer to Procedure 2 section on of your Report Sheet pp. 26–27.

PROCEDURE 3

A Whole Bunch of Electrochemical Cells

In the final part of the research project, you will use a scaled-down, simplified version of electrochemical cells to investigate the relative oxidation – reduction abilities of several promising metals. In order to save time and minimize the use of chemicals, half-cells will be constructed by placing a small piece of metal in contact with filter paper soaked in a metal ion solution (the filter paper serves as a tiny beaker).

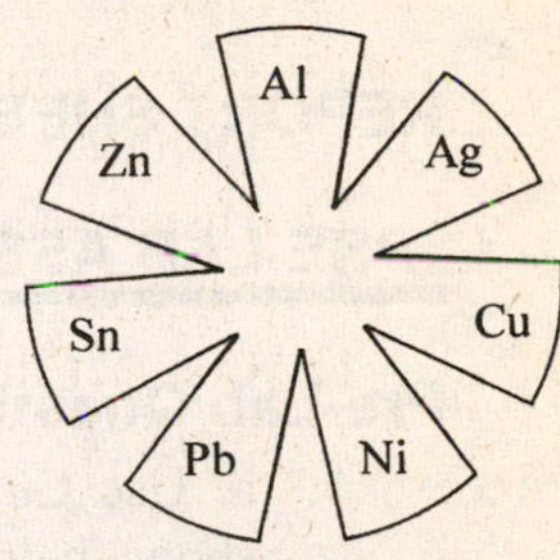

Figure 3.3
Electrochemical "cells"

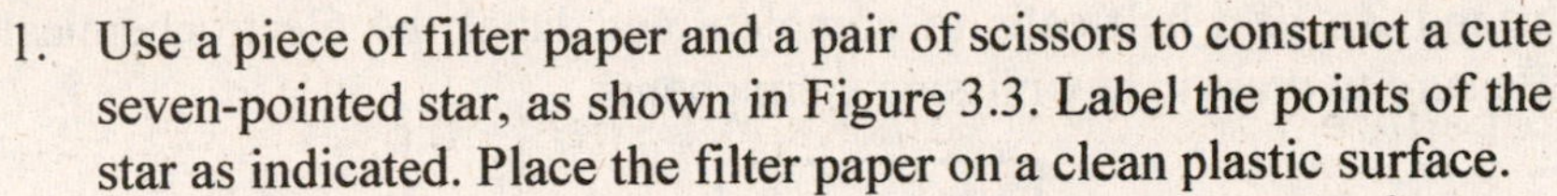

1. Use a piece of filter paper and a pair of scissors to construct a cute seven-pointed star, as shown in Figure 3.3. Label the points of the star as indicated. Place the filter paper on a clean plastic surface.
2. Place 2 to 3 drops of each metal ion solution (e.g., Ag^+, Al^{3+}, or Cu^{2+}) on the appropriate sector on the filter paper. Then place clean pieces of the appropriate metals (e.g. Ag, Al, or Cu) on top of its solution on the appropriate sector.
3. Place 2 to 3 drops of KNO_3 solution in the center of the filter paper so that the KNO_3 joins all the sectors. You may have to rewet the filter paper periodically during the experiment.
4. Use Cu as the reference metal. Touch the black (–) probe to Cu and the red (+) probe to Zn. Record the potential difference (voltage) between the Cu and Zn half-cells.
5. Keep the (–) probe on the Cu reference, and move the (+) probe to Ag. Record the potential difference.
6. Continue in this manner to obtain all the potential differences between Cu and each of the other metals. Note that some readings may be unstable. If so, just try to obtain an approximate voltage.
7. Calculate the reduction potential relative to H^+/H_2 for each metal, and record it.
8. Arrange the metals from highest (most positive) to lowest (most negative) reduction potential. Write the reduction half-reactions, and record the E^o values relative to Cu.
9. When you are done, rinse and dry the metal pieces. Return them to the correct containers. Discard the filter paper as directed. Wash your hands thoroughly!

Refer to Procedure 3 section of your Report Sheet on pp. 27–28.

NAME (print) ______________________________ DATE ______________
LAST FIRST

LABORATORY SECTION ______ PARTNER(S) ______________________________

REPORT SHEET

PRE-LAB ASSIGNMENT

Pre-Lab Questions

1. a. Use Cu^{2+}/Cu and Zn^{2+}/Zn half-cells to complete and label the electrochemical cell shown below. Be sure to draw in any missing components.

V

b. Write the reduction half-reaction and $E°$ ______________________ $E° =$ ________

c. Write the oxidation half-reaction and $E°$ ______________________ $E° =$ ________

d. Write the overall half-reaction and $E°$ ______________________ $E° =$ ________

e. Which substance is the cathode? ________

f. Which substance is the anode? ________

g. Which direction do the electrons flow? ______________________

NAME (print) ____ DATE ____________________
LAST FIRST

LABORATORY SECTION ______PARTNER(S) ____

PROCEDURE 1

1. Record your data for the lead-acid electrochemical cell.

	Submerged Length	Width
+ lead strip		
– lead strip		

	Trial 1	Trial 2	Trial 3
Lead-acid cell Charging time			
Small motor Running time			

2. Calculate the (approximate) total submerged surface area of your lead strips. Show your work.

3. For each of the ERT groups, record the total submerged surface area and maximum motor run time.

Group Name	Electrode Surface Area	Maximum Run Time	Group Name	Electrode Surface Area	Maximum Run Time

NAME (print) ____ DATE ______________________
LAST FIRST

LABORATORY SECTION ______ PARTNER(S) ____

4. What effect did charging time have on how long the motor ran? Is there a charging time after which the motor is fully charged? Explain.

5. What effect does electrode surface area have on how long the motor ran? Explain.

PROCEDURE 2

1. Record your data for the series and parallel electrochemical cells.

Series			Parallel		
Number of Cells	Voltage	Run Time	Number Of Cells	Voltage	Run Time

2. For a given number of lead-acid cells, does series or parallel connection provide the larger voltage? Explain.

NAME (print) ____ DATE ______________
LAST FIRST

LABORATORY SECTION ______ PARTNER(S) ______________________________________

3. For a given number of lead-acid cells, does series or parallel connection provide the longer run time? Explain.

PROCEDURE 3

1. Record your data for the potential difference readings with respect to copper.

Metal						
Voltage						

2. Order the half-reactions from highest reduction potential (most positive voltage relative to Cu) to lowest reduction potential (most negative voltage relative to Cu). Write the reduction half-reaction for each metal and record the measured voltage ($E°$) relative to the Cu^{2+}/Cu reaction. Then add 0.34 V to each of these values to determine the $E°$ relative to the standard H^+/H_2 reaction.

Metal	**Reduction Half-Reaction**	**$E°$ Relative to Cu^{2+}/Cu**	**$E°$ Relative to H^+/H_2**

NAME (print) ______________________________ DATE ____________________
LAST FIRST

LABORATORY SECTION ______ PARTNER(S) ______________________________

3. Based on your results, which two metals would be the best choices for the anode of a battery? Explain.

4. Based on your results, which two metals would be the best choices for the cathode of a battery? Explain.

5. Based on your results, which three metal combinations would yield the highest-voltage battery? Explain.

SOURCES OF ERROR

Once the data have been collected, the possibility of error in the results must be addressed. Please cite at least three random sources of error and briefly explain how each source of error may have affected your data. For a review of error analysis, see *Error Analysis in the Chemistry Laboratory* in the Appendix at the end of this manual.

NAME (print) ________________________________ DATE ____________________
LAST FIRST

LABORATORY SECTION ______ PARTNER(S) ______________________________

CONCLUSION

What factors affected how well the lead-acid battery worked? Specifically, what battery conditions and connections made the motor run the longest? What metal combinations would you recommend for making the best battery? What other things can you conclude or state? State your recommendations and the reasoning behind them in complete sentences. Be sure that your conclusion is complete and concise.

Experiment 4
Copper Reaction Cycle

PRE-LAB ASSIGNMENT

Reading:

1. Read the experiment.

Assignment:

2. Review how to write and balance chemical equations (see Chapter 3 in *Adventures in Chemistry.*)
3. Review the differences between precipitation, acid-base, and oxidation-reduction reactions. (See Chapter 3 in *Adventures in Chemistry.*)

INTRODUCTION

A subsidiary of UniChem International has recently acquired the mining rights to a large section of land that contains the ores of several important metals. UniChem International's geological analysis indicates that the copper-containing ores on the land are readily accessible and particularly rich in copper. UniChem International's economists have studied the global copper market and believe, based on anticipated demand and price, that this is a good time to invest in a copper manufacturing plant.

In the UniChem International copper facility, copper ore will be extracted, refined and purified to produce pure copper metal. UniChem International is also considering the inclusion of a copper reclamation process, which would recover copper metal from waste aqueous solutions containing copper ions. In addition, it is anticipated that a variety of copper compounds also will be manufactured at the plant. If UniChem International decides to proceed with this project, the copper production facility will require a huge investment, with great potential for large profits or losses.

Before committing to this undertaking, UniChem International has assigned its Basic Chemistry Research Division (BCRD) the task of investigating the chemistry of copper and its compounds. Specifically, the BCRD will investigate the ability of copper to react and form different compounds, the reactions needed to chemically convert copper species from one form into another, and the efficiency with which these transformations can be made. (While not necessarily expected, any discoveries of new and useful copper-containing substances would be a bonus!)

As one of the most experienced BCRD chemists, you have been chosen to perform a copper reaction cycle. By using these reactions, you will transform copper metal into a variety of different copper species and eventually return the copper to its elemental state. You also will compare the initial and final amounts of copper metal to determine the overall efficiency of this particular copper reaction cycle.

LEARNING OBJECTIVES

Be able to:

- Perform reactions involving conversions between copper and a variety of its compounds.
- Write balanced chemical equations for the chemical reactions that occur.
- Identify reactions as precipitation, acid–base, or oxidation–reduction.
- Determine percent yield after a series of chemical reactions.

APPARATUS

Chemicals:

- Copper wire
- Concentrated nitric acid (HNO_3)
- 6.0 M sodium hydroxide (NaOH)
- 6.0 M sulfuric acid (H_2SO_4)
- 30-mesh zinc metal
- Methanol

Equipment:

- Stirring rod with rubber policeman
- Porcelain evaporating dish
- Analytical balance (0.001 g accuracy)
- Hot plate

- Are your safety glasses on?
- HNO_3 and H_2SO_4 can burn your skin and clothes.
- Steps 2 and 9 must be done in the fume hood
- NaOH is caustic.
- Methanol is flammable. Do not use a Bunsen burner.

Clean-up:

- Dispose of all excess reagents and waste as directed by your instructor.
- Wash your hands thoroughly with soap and water before you leave the laboratory.

PROCEDURE

1. Cut a piece of pure copper wire that weighs about 0.5 g, and record its exact weight to the thousandth of a gram. Coil the copper wire into a flat spiral and place it in the bottom of a 250-mL beaker.
2. Use a clean, dry 10-mL graduated cylinder to measure 4.0 mL of concentrated nitric acid (HNO_3). In the fume hood, carefully add the HNO_3 to the wire in the beaker. Swirl the resulting solution until all the solid copper has reacted completely. (If some Cu (*s*) is still

left after the reaction stops, you may need to add another 1.0 mL of nitric acid.) Record your observations.

⚠ If you spill any HNO_3, alert your supervisor immediately.

3. After the solid copper has reacted completely, add about 100 mL of distilled water to the solution formed in step 2. Record your observations.
4. Use a clean, dry 25-mL graduated cylinder to measure 15.0 mL of 6.0 M sodium hydroxide (NaOH). Carefully add the NaOH to the solution you made in step 3. Record your observations.

   ⚠ If you spill any NaOH, alert your supervisor immediately.
5. Use a hot plate to carefully heat the solution made in step 4. While stirring continually, heat the solution until it just begins to boil. When no further reaction is observed, remove the beaker from the hot plate and allow the solution to settle. Record your observations.
6. Carefully decant (pour off) the supernatant liquid (the clear liquid above the solid in the beaker), leaving behind the entire solid product.
7. Wash the product left in the beaker after step 6 by adding approximately 150 mL of hot distilled water to the product, stirring to mix thoroughly, allowing it to settle, and decanting the solution.
8. Use a clean, dry 50-mL graduated cylinder to measure 30 mL of 6.0 M sulfuric acid (H_2SO_4). Carefully add the H_2SO_4, with stirring, to the product of step 7. Record your observations.

   ⚠ If you spill any H_2SO_4, alert your supervisor immediately.
9. Obtain about 2 g of 30-mesh zinc metal, and determine its mass to the thousandth of a gram. In the fume hood, add the zinc metal all at once, stirring continually until the solution is colorless.
10. When the reaction is no longer producing any gas, carefully decant the supernatant liquid. Use a rubber policeman to transfer all the solid product to a porcelain evaporating dish. Record your observations.
11. Add about 5 mL of distilled water to the product of step 9, stir and allow it to settle, and then decant. Use another 5 mL of distilled water to repeat this washing process.
12. Add about 5 mL of methanol to the product, stir and allow to settle, and decant again.
13. Place the evaporating dish and its contents on a hot plate on low heat. Carefully heat the solid product until it is completely dry, and then allow the product to cool to room temperature. Describe the appearance of the substance you have produced.
14. Weigh a clean, dry 100-mL beaker to the thousandth of a gram. Carefully transfer as much of the dry product from step 13 as possible to the beaker, and reweigh.

Refer to Copper Cycle Data section of your Report Sheet on p. 34.

NAME (print) ________________________________ DATE ____________
LAST FIRST

LABORATORY SECTION ______ PARTNER(S) ________________________

REPORT SHEET

Copper Reaction Cycle Data

Weight of copper wire ____________

Step 2 observations: ________________________

____

____

Step 3 observations: ________________________

____

Step 4 observations: ________________________

____

Step 5 observations: ________________________

____

Step 8 observations: ________________________

____

Steps 9–10 observations: ________________________

____

Step 13 observations: ________________________

____

Weight of beaker + product ____________

Weight of empty beaker ____________

NAME (print) ____ DATE ____________________
LAST FIRST

LABORATORY SECTION ______ PARTNER(S) ____

DATA ANALYSIS

Copper Cycle Reactions and Calculations

1. In step 2, the copper metal reacts with nitric acid to form a solution containing aqueous $Cu(NO_3)_2$, the gas NO_2 (a component of smog) and water. (The beautiful blue color of the solution is due to the Cu^{2+} ion.)

   a. Write the balanced chemical equation for this reaction.

   b. What type of reaction is this (i.e., acid–base, precipitation, oxidation–reduction)?

2. In step 4, the copper(II) nitrate (produced earlier in step 2) reacts with the added sodium hydroxide to produce solid copper(II) hydroxide and aqueous sodium nitrate.

   a. Write the balanced chemical equation for this reaction.

   b. What type of reaction is this?

3. The heating in step 5 produces copper(II) oxide and water from the copper(II) hydroxide (formed previously in step 4).

   a. Write the balanced chemical equation for this reaction.

   b. What type of reaction is this?

4. In step 8, the copper(II) oxide reacts with the sulfuric acid to produce an aqueous solution of copper(II) sulfate and water.

   a. Write the balanced chemical equation for this reaction.

   b. Which specific species causes the color of this solution?

   c. What type of reaction is this?

NAME (print) ______________________________________ DATE ______________
LAST FIRST

LABORATORY SECTION ______ PARTNER(S) ______________________________________

5. In steps 9–10, zinc metal reacts with aqueous copper(II) sulfate to produce solid copper and aqueous zinc sulfate.

   a. Write the balanced chemical equation for this reaction.

   b. What type of reaction is this?

6. When the zinc was added in step 9, it also reacted with leftover acid (H^+) from step 8, producing a flammable gas and Zn^{2+} ion.

   a. Write the balanced chemical equation for this reaction equation.

   b. What type of reaction is this?

7. Use the weight of the empty 100-mL beaker and the beaker plus dry product from step 15 to determine the mass of copper recovered. Show your calculations.

8. Use the original mass of copper wire and the final mass of copper product to determine the percent of copper recovered. Show your setup and calculations.

$$\text{Percent of copper recovered} = \left(\frac{\text{Weight of copper recovered}}{\text{Original weight of copper wire}}\right) \cdot 100\,\%$$

NAME (print) ____ DATE ____________________
LAST FIRST

LABORATORY SECTION ______ PARTNER(S) ____

SOURCES OF ERROR

Once the data have been collected, the possibility of error in the results must be addressed. Please cite at least two random sources of error and briefly explain how each source of error may have affected your data. For a review of error analysis, see *Error Analysis in the Chemistry Laboratory* in the Appendix at the end of this manual.

CONCLUSION

Based on your analysis, how successful were you in converting copper metal into a variety of copper compounds and then back into copper? Rate the overall efficiency of conversion. Answer the question in complete sentences. Be sure that your conclusion is complete and concise.

Experiment 5

Analysis of Antacids

PRE-LAB ASSIGNMENT

Reading:

1. Read the experiment.

Assignment:

2. Find the active ingredient in two commercial antacids. Please use an antacid different from milk of magnesia.
3. Provide full references for your sources.
4. Write two separate balanced chemical equations between your active ingredients and the hydrogen ion (H^+.) (See Chapter 3 in *Adventures in Chemistry.*)

INTRODUCTION

Digestion of the food we eat begins way before the food gets to our stomachs. However, most of the digestion takes places in our stomach and is a result of our gastric fluids interacting with the foods we eat. Gastric juices, which mainly consist of hydrochloric acid, are very acidic with a pH of about 2. The presence of stomach acid is very important component of digestion. One of the roles of stomach acid is to convert inactive enzymes, needed for digestion, into active enzymes. One such inactive enzyme is pepsinogen, which is converted to the active enzyme pepsin. Too much of a good thing can turn into a bad thing. When the stomach produces too much acid, this may result in heartburn or acid reflux, which may lead to the formation of ulcers. Stresses such as too much food, pregnancy, and constipation may trigger a heartburn attack.

To relieve the symptoms of heartburn we may take an antacid. Antacids, regardless of the manufacturer, have one role, to neutralize excess stomach acid. All commercial antacids contain a weak basic ingredient, such as hydroxides and carbonates that reacts with the hydrogen ion in the stomach via an acid–base reaction. Magnesium hydroxide, the major component of Milk of Magnesia, reacts with the hydrogen ion as follows:

$$Mg(OH)_2 + 2\,H^+ \rightarrow Mg^{2+} + 2H_2O$$

How effective an antacid will be depends on how much acid it can neutralize. Manufacturers of the different types of antacids all claim that their antacid is the best at neutralizing excess acid in the stomach and thus relieves heartburn.

The American Medical Association (AMA) wants to put its seal of approval on the best antacid on the market today. The antacid that will bear the AMA's gold seal of approval will be the one that neutralizes the most stomach acid per tablet. The AMA has hired the Drug Characterization Division (DCD) of UniChem International to determine which of the major manufacturers of commercially available antacids that produces a product that is most effective at neutralizing excess stomach acid and the winner of the gold seal.

In this experiment, you and a fellow DCD analyst will determine which antacid is most effective at neutralizing excess stomach acid by using a concept known as **strong acid–strong base titration**. Titration is a very useful technique in analytical chemistry because it is a way of determining the concentration of an unknown acid or base. Reviewing acid–base chemistry from Chapter 4 of *Adventures in Chemistry*, we know that sodium hydroxide will neutralize hydrochloric acid to produce water and sodium chloride by the following reaction:

$$NaOH + HCl \rightarrow H_2O + NaCl$$

Notice from this balanced chemical equation, that every mole of the base present will neutralize one mole of the acid and produce one mole of water and one mole of sodium chloride. Using this relationship, the concentration of acid present in the solution can be determined using the concentration and volume of the basic solution and the volume of the acidic solution.

$$\text{Concentration}_{\text{base}} \times \text{volume}_{\text{base}} = \text{concentration}_{\text{acid}} \times \text{volume}_{\text{acid}}$$

The acidic solution will contain an indicator. An **indicator** is a substance (either an acid or a base) that is one color at the beginning of the titration and changes colors toward the end of the titration. The point at which the indicator changes colors is called the **endpoint** of the titration. As the base is added to the acid, the pH of the solution will change until a complete neutralization has taken place, which is when the indicator changes color. The indicator you will be using is phenolphthalein. Phenolphthalein is colorless under acidic conditions and changes to pink under basic conditions ($pH > 7$).

In this experiment, you and your partner will determine the amount of HCl neutralized by antacids manufactured by different companies to determine which antacid is more effective. We will be using a type of titration called **backtitration**. Backtitration is used because antacids are not completely soluble in water but are soluble in excess HCl. Back titration is very similar to normal titration, where the major difference is that a known excess of one reagent (HCl) is added to the solution titrated (antacid in HCl). The solution then is titrated back to determine how much of the known excess HCl reagent remains.

Each antacid tablet will be dissolved in an excess amount of HCl with a known concentration and volume of HCl. Excess HCl is necessary because it prevents the possibility of a buffer system establishing and it helps to dissolve the antacid. A **buffer system** is a pair of acids and bases that maintains pH. The active component of the antacid (weak base) will neutralize some of the HCl. How many moles of HCl remains will be determined by titrating with a standard sodium hydroxide solution until the endpoint is reach, where the color of the indicator changes from colorless to light pink. The moles of HCl remaining in solution can be calculated using the concentration and the volume of the base and remembering that one mole of base neutralizes one mole of acid (see sample calculations below). Once the antacid has neutralized all the acid it is able to, moles of acid will remain. How effective the antacid is will depend on the number of moles left in the flask. The most effective antacid will leave the fewest number of moles of acid in the flask.

Since all antacids are not created equal, we need to have a relative way of comparing the different antacids. We will determine which antacid is the most effective by calculating the grams of HCl neutralized per gram of antacid. All antacid data will be entered into a spreadsheet for comparison. Once all the data have been entered into the spreadsheet, the analyst from the DCD at UniChem International will be able to determine which antacid is most effective at neutralizing excess stomach acid. This information will be passed on to the AMA to indicate which manufacturer of antacids should be awarded AMA's gold seal of approval.

Sample Calculation:

Mass of antacid:	0.458g
Concentration of HCl used:	0.1958 *M*
Volume of HCl used:	50.00 mL
Initial moles of HCl used:	0.009790 mol
molHCl = (0.05000 L) × (0.1958 mol/L).	
Concentration of NaOH used:	0.2141 *M*
Final burette reading:	29.82 mL
Initial burette reading:	1.420 mL
Volume of NaOH used: (29.82mL – 1.420mL)	28.40 mL

Remember from Chapter 4 that **molarity** is defined as moles per liter. Rearranging, we find that moles equals molarity times liters:

mol_{NaOH} = (volume of NaOH from titration) x (molarity of NaOH)
= (0.02840 L) × (0.2141 mol/L)
= 0.006080 mol_{NaOH} = mol of HCl left over in flask after neutralization.

Find moles of HCl neutralized by the antacid by taking the difference of how much acid was present initially versus how much acid is left over in the flask.

$$\text{mol HCl}_{\text{neutralized by antacid}} = \text{mol HCl}_{\text{initial}} - \text{mol HCl}_{\text{leftover}}$$
$$= 9.790 \times 10^{-3}\text{ mol} - 6.080 \times 10^{-3}\text{ mol}$$
$$= 3.710 \times 10^{-3}\text{ mol}$$

Moles of HCl neutralized by the antacid must equal the initial moles present in the antacid tablet.

We are trying to find the mass of HCl neutralized per gram of antacid. Therefore, we need to convert moles of HCl neutralized to grams of HCl neutralized. To do this we need to use both the molarity and density of the HCl solution. Using the molarity we can convert moles of HCl to milliliters. Molarity is equal to moles per volume. Volume can be determined by dividing the moles by the molarity:

$$\text{mL of HCl} = 3.710 \times 10^{-3}\text{ mol HCl} \times \frac{1.000\text{L HCl}}{0.1\text{ mol HCl}} \times \frac{1000\text{ mL}}{1.000\text{L}} = 37.10\text{ mL HCl}$$

Using the density of HCl we can convert milliliters of HCl to grams of HCl neutralized. The density of the HCl solution is assumed to be the same as water, or 1.00 g/mL. Therefore, milliliters of HCl are equal to grams of HCl (37.10 mL HCl = 37.10 g HCl). To determine the mass of HCl neutralized per grams of antacid, divide the mass of the HCl solution by the mass of the antacid tablet as shown below:

$$\frac{37.10\ \text{g HCl solution}}{0.458\ \text{g of antacid}} = 81.0\ \text{g HCl/g antacid}$$

LEARNING OBJECTIVES

Be able to:

- Perform acid–base titrations.
- Determine the amount of acid neutralized by antacids sold by different manufacturers.
- Determine which manufacturer sells the most effective antacid tablet.
- Report findings to the AMA.

APPARATUS

Chemicals:

- Various commercially available antacids
- ~200 mL of 0.100 M standardized HCl solution
- ~300 mL of 0.100 M standardized NaOH solution
- Phenolphthalein indicator in small dropper bottles

Equipment:

- Burette and burette clamp
- Small plastic funnel
- Ring stand
- 50.0-mL Volumetric pipettes
- 1–2 Pipette bulbs
- Analytical balance
- Hot plate with stirring
- Magnetic stir bar
- Magnetic stir bar remover
- Mortar and pestle
- 3–4 Plastic weighing dishes
- Roll of colored tape for labeling
- Computer with Excel® spreadsheet for class data.

- Are your safety glasses on?
- HCl and NaOH can burn your skin and clothes.
- Phenolphthalein indicator may stain clothes and skin.
- Please wear gloves when handling these reagents.
- Please do not eat any of the antacids.

Clean-up: 

- No special clean-up is required.
- All solutions can go down the drain with water.
- Wash your hands thorough with soap and water before you leave the laboratory.

PROCEDURE 1

Antacid Information

1. Record the name of the antacid, the manufacturer, the active ingredient and the amount of active ingredient of all the various types of commercially available antacids on your Report Sheet.
2. Record the exact concentration of the standardized HCl and NaOH solutions on your Report Sheet.
3. Select one type of the antacids available for testing and record the name of the antacid on your Report Sheet.

Refer to Procedure 1 Data Table of your Report Sheet on p. 45.

PROCEDURE 2

Antacid Analyses

1. Perform three titration trials for each type antacid you test.
2. Obtain one antacid from the container.
3. Using a mortar and pestle, crush the antacid until it is a fine powder.
4. Obtain three clean, dry 250-mL Erlenmeyer flask.
5. Label the 250-mL Erlenmeyer flasks as Trial 1, Trial 2, and Trial 3.
6. Pre-weigh the 250-mL Erlenmeyer flask labeled as Trial 1.
7. Record this weight on your Report Sheet.
8. Be sure to record all figures from the balance.
9. Transfer the crushed antacid from the mortar to the preweighed flask.
10. Re-weigh the flask and antacid using the same balance.
11. Record this weight on your Report Sheet. Be sure to record all the digits.
12. Determine the weight of the antacid by difference (Subtract the weight of the empty flask from the weight of the flask and antacid).
13. Record the weight of the antacid on your Report Sheet.
14. Following your supervisor's instructions, add 50.00 mL of the standardized HCl solution to the flask using a volumetric pipette.
15. Swirl to dissolve the antacid as much as possible.
    - ➢ The entire tablet may not dissolve because inert binders and starches are used to hold the tablet together.
16. Slowly slide a magnetic stir bar down the side of the Erlenmeyer flask, and place the flask on the hot plate.
17. Heat the solution to a gentle boil with stirring and continue to boil for 2–3 minutes.
    - ➢ This is necessary to expel carbon dioxide gas that may have formed during the neutralization process.

18. At the end of the heating process, carefully remove the hot flask from the hot plate using your beaker tongs.
19. Add –8 drops of phenolphthalein indicator.
    - ➢ Phenolphthalein is colorless under acidic conditions and pink under basic conditions.
20. If the solution is pink, add another 25.0 mL of the 0.1 M HCl solution and heat until boiling. Record the new volume of HCl on your Report Sheet.
21. Remove the flask from the hot plate and allow the solution to cool to room temperature while you are assembling the titration apparatus.
22. Assemble a titration apparatus like the one in Figure 5.1.
23. Obtain the standardized sodium hydroxide, and be sure the exact concentration has been recorded on your Report Sheet.
24. Add 10 mL of the NaOH solution to the burette and rinse the burette with the solution. The solution may be easier to pour into your burette if you use a small plastic funnel.
25. Open the stopcock until air bubbles move through the stopcock and allow the NaOH solution to drain into a beaker.
26. Close the stopcock and fill the burette with NaOH slightly above the 0.00-mL mark.
27. Open the stopcock to allow NaOH to drain into the beaker until the level of NaOH is as close to the 0.00-mL mark. It is okay to be below the mark but not above the mark.
    - ➢ This is the initial burette reading.
28. Record the initial reading on your report Sheet.
29. Return the flask to the stirrer and start stirring.
30. Make sure that the heat is off.

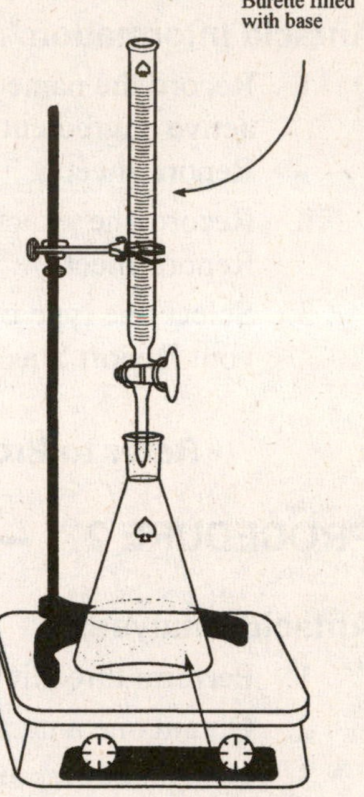

Figure 5.1 Titration apparatus

31. Titrate the antacid sample until a pale pink color lasts for at least 30 seconds.
32. Record the final buret reading on your Report Sheet.
33. Refill the burette with NaOH.
34. Repeat the procedure with the remaining two antacid samples.
35. Enter your data for your antacid(s) on the DCD spreadsheet. You will be given a copy of the spreadsheet in order to compare the antacids tested.

Refer to Procedure 2 section of your Report Sheet on p. 46.

DATA ANALYSIS

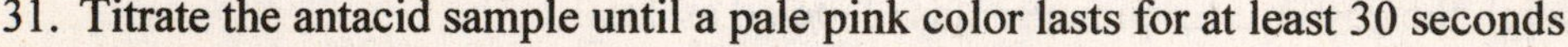

1. Calculate the amount of HCl neutralized per gram of antacid tablet tested.
2. Record your values on your Report Sheet.
3. Using the DCD spreadsheet, record the amount of HCl neutralized per gram of antacid tablet by other DCD agents.
4. Record these values on your Report Sheet.
5. Which manufacturer of which antacid should be awarded the gold seal of approval from the AMA?

Refer to Data Analysis Data Tables III and IV of your Report Sheet on pp. 46–47.

NAME (print) ____ DATE______________
LAST FIRST

LABORATORY SECTION ________ PARTNER(S) ______________________________________

REPORT SHEET

PRE-LAB ASSIGNMENT

Pre-Lab Questions

1. Find the active ingredient in two commercial antacids. Please use an antacid different from milk of magnesia.

2. Write two separate balanced chemical equations between your active ingredients and the hydrogen ion (H^+.)

PROCEDURE 1

Data Table I: Record the following information for all the available antacids.

Brand of Antacid	Active Ingredient(s)	Amount of Active Ingredient(s)	Manufacturer

Molarity of HCl solution: ______________

Volume of HCl added: ______________

Molarity of NaOH solution: ______________

Antacid used: ______________

NAME (print)____ DATE____________________
LAST FIRST

LABORATORY SECTION ______ PARTNER(S)____________________________________

PROCEDURE 2

Data Table II: Record mass of antacid and titration data.

Trial	Mass of Flask (with Antacid)	Mass of Flask (Empty)	Mass of Antacid	Final Burette Reading	Initial Burette Reading	Volume of NaOH Used
# 1						
# 2						
# 3						

DATA ANALYSIS

Data Table III: Calculations.

Calculations	Trial #1	Trial #2	Trial #3
Mass of antacid			
Volume of NaOH used			
Moles of HCl (initial amount)			
Moles of HCl remaining (after neutralizing)			
Moles of HCl neutralized			
Milliliters of HCl Neutralized			
Grams of HCl Neutralized			
Grams of HCl neutralized per gram of antacid			

NAME (print) ____ DATE ____________________
LAST FIRST

LABORATORY SECTION ______ PARTNER(S) ____

Sample Calculations: Show all calculations for one trial.

Data Table IV: Summary of antacid effectiveness

Brand of Antacid Tested	Mass of HCl Neutralized per Gram of Antacid	Manufacturer

NAME (print) ____ DATE ____________________
LAST FIRST
LABORATORY SECTION ______ PARTNER(S) ____

SOURCES OF ERROR

Once the data have been collected, the possibility of error in the results must be addressed. Please cite at least three random sources of error and briefly explain how each source of error may have affected your data. For a review of error analysis, see *Error Analysis in the Chemistry Laboratory* in the Appendix at the end of this manual.

POST-LAB QUESTIONS

1. When an antacid is taken to relieve heartburn a gas is produced during the reaction. The gas produced is carbon dioxide. Name one antacid tested that will produce CO_2 during the reaction.

2. Using the antacid from question 1, write a balanced chemical equation between the active ingredient and hydrochloric acid that shows the formation of CO_2.

NAME (print) ______________________________ DATE ______________
LAST FIRST

LABORATORY SECTION ______ PARTNER(S) ______________________________

3. You have just been hired by a pharmaceutical company that wants to design a new antacid using sodium hydroxide as the active ingredient. One reason the company wants to use sodium hydroxide is because of its solubility and the other reason it because of its effectiveness in neutralizing acids. Give one reason why sodium hydroxide should not be used as the active ingredient in an antacid.

CONCLUSION

Based on your findings and those of your fellow Drug Characterization Division agents, which antacid is the most effective at neutralizing excess stomach acid? What data can be used to support your findings to the American Medical Association? What other things can you conclude or state? State your conclusions and the reasoning behind them in complete sentences. Be sure that your conclusion is complete and concise.

Experiment 6

Synthesis of Soap—Saponification

PRE-LAB ASSIGNMENT

Reading:

1. Read the experiment.

Research:

2. Research the chemistry of soap in both synthesis and function (Chapter 4 of *Adventures in Chemistry* provides some relevant information).

Question:

3. Answer the Pre-Lab Questions.

INTRODUCTION

Since approximately 500 BC, it has been known that heating a mixture of animal fat and wood ashes produces a thick, creamy material with many interesting and useful properties. After the initial discovery (probably serendipitously as grease from cooking meat spilled into the fire), several methods were developed to make this substance, which we call soap, purposefully.

We use soap (or synthetic detergent) to clean clothing, dishes, floors, and much more. Using soap when we bathe removes dirt and body oils from our skin, keeping the pores from clogging. Washing a cut or other wound with soap helps to kill bacteria and keep the wound from becoming infected. In fact, recent research has shown that regular soap does this just as well as soap to which antibacterial ingredients have been added.

The reaction to make soap, known as **saponification**, is technically a base-promoted hydrolysis of the ester linkages in fats or oils. A typical saponification reaction is shown below:

$$\begin{array}{l} CH_2\text{-}O\text{-}\overset{O}{\overset{\|}{C}}\text{-}(CH_2)_{16}CH_3 \\ | \\ CH\text{-}O\text{-}\overset{O}{\overset{\|}{C}}\text{-}(CH_2)_{16}CH_3 \\ | \\ CH_2\text{-}O\text{-}\overset{O}{\overset{\|}{C}}\text{-}(CH_2)_{16}CH_3 \end{array} + 3\,NaOH \xrightarrow[\text{Heat}]{H_2O} \begin{array}{l} CH_2\text{-}OH \\ | \\ CH\text{-}OH \\ | \\ CH_2\text{-}OH \end{array} + 3\,Na^{+}\ {}^{-}O\text{-}\overset{O}{\overset{\|}{C}}\text{-}(CH_2)_{16}CH_3$$

Tristearin Glycerol Sodium stearate

Soap is made by heating a fat or oil (e.g., vegetable oil, coconut oil or animal fat) and an alkali (usually sodium hydroxide, also known as **lye**). In the **kettle method**, these ingredients are mixed in a large heated vat (the "kettle") for several hours. Then a concentrated salt solution is added to the soap, and the less dense soap floats to the top and is removed and purified. Now, most soap is made by a faster method known as **continuous processing**. In this method, hot pressurized water and hot fat are combined in a long tube called a **hydrolyzer**. The water, rather than lye, serves to hydrolyze the triglyceride and produce soap.

Soap is an effective cleaning agent because of its molecular structure. One end of a soap molecule is polar, and since it strongly attracts water, it is called **hydrophilic**. The other end of the molecule is nonpolar and attracts other nonpolar substances but repels water and thus is called **hydrophobic**. When soap molecules encounter dirt or grease, they form spherical micelles with the hydrophobic ends surrounding the grease and the hydrophilic ends pointing outward. Added water then will carry the micelles, and the grease they contain, away from the dirty item and leave it clean.

The top synthetic chemists in the UniChem International's Cleaning Products Division (CPD) recently have developed a top-secret soap recipe. They believe that their formulation will provide an extremely effective cleaning substance and make UniChem International a major player in the cleaning products field. However, before UniChem begins soap production, adequate testing of this top-secret soap recipe is necessary.

This task has been assigned to the Product Testing Department. You and your fellow product testers will use the top-secret formula to make small quantities of the soap, assess the process and its yield, and test the properties of the soap. In addition, you will investigate modifications to the process to see if you can develop a soap with improved characteristics, such as cleaning capacity, feel, smell, moisturizing ability, etc.

LEARNING OBJECTIVES

Be able to:

- Synthesize soap.
- Analyze the characteristics of the soap you create.
- Determine what makes a "better" soap.

APPARATUS

Chemicals:

- Various oils (e.g., corn oil, olive oil, coconut oil, etc.)
- Various fats (e.g., lard, Crisco, etc.)
- Ethanol
- 6.0 M sodium hydroxide (NaOH)
- Saturated sodium chloride (NaCl) solution
- Mineral oil
- 3% $CaCl_2$ solution
- 3% $FeCl_3$ solution
- Commercial detergent solution

Equipment:

- Hotplate
- Boiling chips
- Büchner funnel
- Filter flask with sidearm
- Vacuum lines or suction nozzles on water faucets
- Vacuum tubing
- Filter paper
- Ring stand and clamp
- pH paper

- Are your safety glasses on?
- NaOH is caustic.
- Ethanol is flammable. Do not use a Bunsen burner.

Clean-up:

- Dispose of all excess reagents and waste as directed by your instructor.
- Wash your hands thoroughly with soap and water before you leave the laboratory.

PROCEDURE 1

Making Soap:

1. First, create a hot-water bath. Add tap water to a 600-mL beaker until it reaches the 200-mL line. Place a few boiling chips in the beaker. Place the beaker on a hot plate and turn on the heat. While you are waiting for the water to boil, continue to step 2.
2. Use a graduated cylinder to measure about 15 mL of vegetable oil. Pour the oil into a 250-mL Erlenmeyer flask.
3. Use a graduated cylinder to measure about 15 mL of ethanol. Add the ethanol to the Erlenmeyer flask. Use a glass stirring rod to mix the liquids.
4. Use a graduated cylinder to measure about 12 mL of 6.0 M NaOH solution. Add the NaOH to the Erlenmeyer flask. Stir to mix the liquids.
   ⚠If you spill NaOH, please alert your supervisor immediately.
5. When the water in the hot-water bath has begun to boil, place the Erlenmeyer flask in the beaker. Refer to Figure 6.1.
6. Stir the mixture in the flask continuously. Make sure that the water is at a gentle boil. Watch to ensure that the solution does not boil over and that no water splashes into the solution.

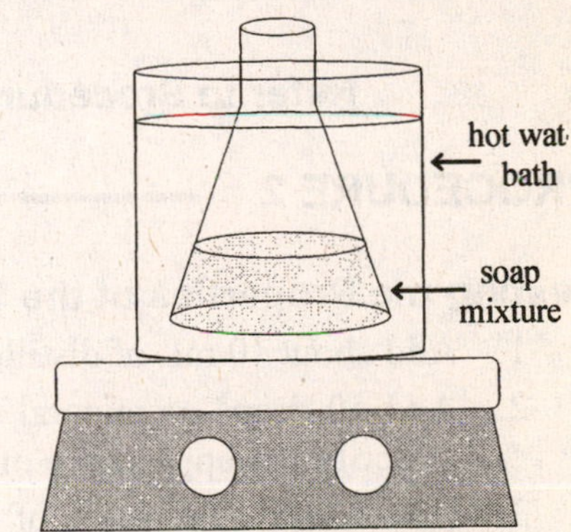

Figure 6.1
Saponification setup

7. Continue to heat the solution, with constant stirring, until the ethanol odor disappears. This indicates that the reaction is complete.
8. Now create an ice-water bath. Add crushed ice to a 600-mL beaker until it is about half-full. Then add enough water to the ice to make about 200 mL of ice-water mixture.
9. Use tongs to move the Erlenmeyer flask from the hot-water bath to the ice-water bath. At this point, the flask contains soap, as well as glycerol and excess NaOH.
10. Add about 120 mL of saturated NaCl solution to the soap mixture while stirring vigorously. This will increase the density of the aqueous solution, which then will allow the (less dense) soap to float to the top of the solution.
11. Use a Büchner funnel, a piece of filter paper, a filter flask with sidearm, a ring stand and clamp, and some suction tubing to set up for filtration as shown in Figure 6.2.

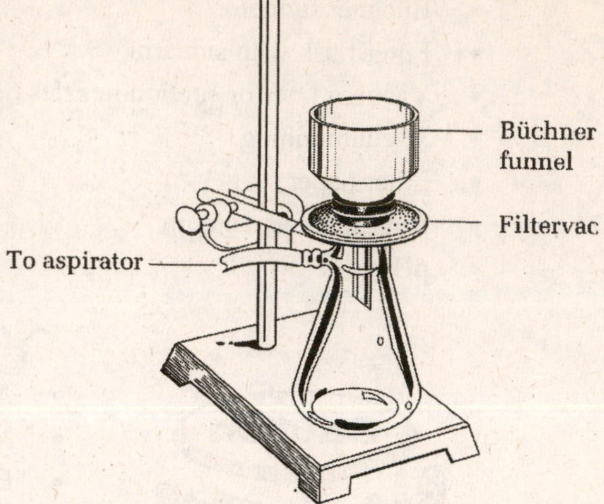

Figure 6.2 Filtering the soap

12. Place a piece of filter paper in the Büchner funnel, as directed. Use a squirt bottle of distilled water to carefully wet the filter paper so that it lies flat in the funnel and covers all the holes.
13. As directed by your supervisor, turn on the suction. Carefully transfer the soap mixture into the top of the Büchner funnel and filter out the soap. Use a rubber policeman to transfer as much of the soap as possible.
14. Leave the suction on, and wash the soap by slowly pouring about 25 mL of saturated NaCl solution over the soap.
15. Repeat this washing procedure with a second 25-mL portion of saturated NaCl solution.
16. Repeat this washing procedure with about 15 mL of ice-cold distilled water.
17. Describe the appearance of your final soap product (i.e., phase, color, texture, etc.).
18. As directed by your supervisor, either repeat this soap-making procedure with a different oil or fat or share your results with other product testers who used a different fat or oil to make their soap.

Refer to Procedure 1 section of your Report Sheet on p. 57.

PROCEDURE 2

Testing the Properties of the Soap:

1. Add about 10 mL of distilled water to a medium test tube.
2. Add 10 drops of mineral oil to the test tube. Cork the test tube and shake vigorously. Describe the appearance of the water/mineral oil mixture.
3. Now add a small piece of soap to the water–mineral oil test tube. Cork the test tube and shake vigorously as before. Describe the effect the soap has on the mineral oil/water mixture.
4. Combine about 4 mL of soap and 50 mL of distilled water in a 150-mL beaker. While stirring, carefully warm the beaker and contents until the soap is dissolved.

5. Label four medium test tubes (#1, #2, #3 and #4). Pour about 10 mL of soap-water solution into each test tube.
6. Use a piece of pH paper to determine the pH of the soap solution in test tube #1. Record the color and pH value.
7. Add 5 drops of $CaCl_2$ solution to test tube #2. Stopper the test tube and shake to mix. Record your observations.
8. Add 5 drops of $FeCl_3$ solution to test tube #3. Stopper the test tube and shake to mix. Record your observations.
9. Add 5 drops of tap water to test tube #4. Stopper the test tube and shake to mix. Record your observations.
10. Combine about 4 mL of detergent and 50 mL of distilled water in a 150-mL beaker. While stirring, carefully warm the beaker and contents until the soap is dissolved.
11. Repeat steps 5–9 but using the detergent solution instead of the soap solution.

Refer to Procedure 2 section of your Report Sheet on pp. 57–58.

NAME (print) ______________________________ DATE ______________
LAST FIRST

LABORATORY SECTION ______ PARTNER(S) ______________________________

REPORT SHEET

PRE-LAB ASSIGNMENT

Pre-Lab Questions

1. Write the equation for the saponification reaction that produces sodium linoleate, $CH_3(CH_2)_4CH{=}CHCH_2CH{=}CH(CH_2)_7CO_2^-Na^+$. Be sure to draw all molecular structures in condensed form. (You may want to refer to the saponification reaction that produces sodium stearate shown in the Introduction.)

2. Using appropriate diagrams at a molecular level, describe how soap functions to remove grease or oil.

3. Using appropriate diagrams at a molecular level, describe how soap scum forms.

4. Describe the difference between soap and detergent.

NAME (print) ______________________________ DATE ______________
LAST FIRST

LABORATORY SECTION ______ PARTNER(S) ______________________________

PROCEDURE 1

1. Record your observations of the various soaps produced.

Type of Fat or Oil Used	Description of Soap

PROCEDURE 2

1. Record your observations on the emulsifying effects of soap.

   a. Describe the mixture formed by mineral oil and water.

   b. Describe what happens when soap is added to the mineral oil–water mixture.

NAME (print) ______________________________ DATE ____________
LAST FIRST

LABORATORY SECTION ______ PARTNER(S) ____________________

2. Record your observations of the soap testing.

Test Tube #	Test Performed	Observations

3. Record your observations of the detergent testing.

Test Tube #	Test Performed	Observations

NAME (print) ______________________________ DATE ______________
LAST FIRST

LABORATORY SECTION ______ PARTNER(S) ______________________________

POST-LAB QUESTIONS

1. Which fat or oil produced the "best" soap? Explain your reasoning.

2. Are the soaps you made using the top-secret recipe suitable for production and marketing at this point? If not, what would you do to improve your soap product? Make specific suggestions regarding additives and the qualities they should provide.

3. If you do not want to produce soap scum, is it better to use soap or detergent? Explain based on the results of your soap and detergent testing.

CONCLUSION

Based on your product testing, summarize your findings regarding the top-secret soap-making recipe you were given and the quality and effectiveness of the soap produced. State your recommendations and the reasoning behind them in sentence format. Be sure that your conclusion is complete and concise.

Experiment 7
Distillation

PRE-LAB ASSIGNMENT

Reading:

1. Read the experiment.
2. Review the types and relative strengths of intermolecular forces (refer to Chapter 4 of *Adventures in Chemistry*).
3. Review the structures and intermolecular attractions of simple hydrocarbons (refer to Chapter 7 of *Adventures in Chemistry*).

Assignment:

4. Research how a fractionating tower works to separate the substances in crude oil.

Questions:

5. Answer the Pre-Lab Questions.

INTRODUCTION

Distillation is a chemical process for separating individual substances from a mixture. In a typical distillation, a solution is heated in a boiler. The substance that boils at the lowest temperature is the first to vaporize. Once it becomes a gas, the substance enters a condenser, is cooled, and condenses to liquid. This liquid, called the **distillate**, is then collected in a receiver. Using this method, a liquid can be separated from a solid, for example, to produce freshwater from saltwater. If a solution consists of two liquids, fractional distillation will produce a distillate that is enriched in the more volatile liquid, for example in making distilled liquors. Rectification, which is similar to fractional distillation, is used to separate many different substances in a fractionating column, such as in refining crude oil. Other types of distillation include destructive distillation, in which the heating is sufficient to change the substances chemically.

Several divisions of UniChem International are considering expanding. The Petroleum Division is looking into building an oil refinery in western Africa, near newly discovered petroleum reserves, to separate gasoline and kerosene from crude oil. The Libations Division is considering starting a distillery in Scotland to make a fine brandy. And the Water Division is interested in constructing a plant to produce pure water from polluted saltwater in southern California. UniChem International has given your team of distillation experts the task of investigating these potential projects and reporting how well they work. Your findings, combined with those of other analysts investigating the financial and environmental aspects of each project, will be used to make the final determination as to whether to invest in any of these proposals.

LEARNING OBJECTIVES

Be able to:

- Set up a distillation apparatus and use it to separate substances from mixtures.
- Explain how distillation works in terms of the intermolecular attractions.
- Order different compounds in terms of their relative boiling points.

APPARATUS

Chemicals:

- Crude oil
- Wine
- Polluted saltwater sample
- 1.0 M NaCl solution
- 1.0 M $AgNO_3$ solution

Equipment:

- Heating mantle and controller
- Round-bottom flask
- Condenser
- Distillation head
- Thermometer and seal
- Adapter
- Grease
- Keck clamps
- Latex tubing
- Boiling chips
- Hydrometer (or 10- or 25-mL graduated cylinder and analytical balance)

- Are your safety glasses on?
- Do the crude oil distillation in the fume hood.
- The distillation apparatus reaches high temperatures.
- Burn the gasoline and kerosene in the fume hood.
- Do not sample the wine or wine distillation product.

Clean-up:

- Dispose of all excess reagents and waste as directed by your supervisor.
- When they have cooled, thoroughly clean all distillation apparatus pieces, as directed, and return them to the designated area.
- Wash your hands thoroughly with soap and water before you leave the laboratory.

PROCEDURE 1

Setting Up the Distillation Apparatus

1. Set up your assigned distillation apparatus by following the directions below and referring to Figure 7.1.
2. Use an iron ring and ring stand to support a heating mantle.
3. Place a round-bottom flask on the heating mantle, and gently but firmly clamp the neck of the flask to the ring stand.
4. Carefully pour enough of your assigned solution to approximately half fill the round-bottom flask (your still pot), and add two or three boiling chips. (Be sure that the pot is no more than two-thirds full, or the surface area will be too small for rapid evaporation.)

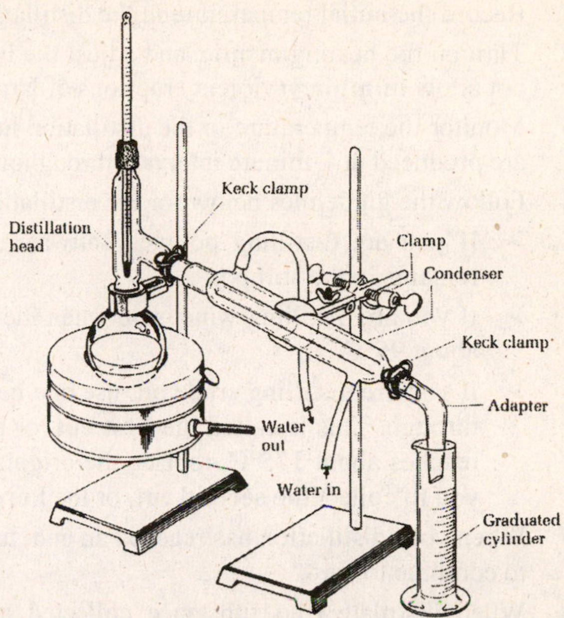

Figure 7.1 Simple distillation apparatus

5. Place a small amount of grease around the ground glass of the bottom male joint of the distillation head, and carefully but firmly press the sidearm into the top of the flask. Do not force the connection, but be sure that it is solid.
6. In the same way, attach the condenser to the sidearm. Use a second ring stand and clamp to support the condenser. Be sure that the clamp holds the condenser securely but does not break it. Use a Keck clamp to secure the condenser to the distillation head.
7. Center the thermometer in the distillation head, and line up the top of the thermometer bulb with the bottom of the sidearm.
8. Attach an adapter to the condenser. Use a Keck clamp to secure the condenser to the adapter.
9. Connect one piece of latex tubing from the faucet to the lower nipple of the condenser. (It is important that the water flow in the bottom of the condenser and out the top.) Be sure that the connections are secure, and that the tubing is neither kinked nor stretched too tightly.
10. Connect a second piece of tubing to the upper nipple of the condenser. Run the tubing from the condenser to the sink. Be sure that the output hose extends completely into a drain.
11. Place a clean, dry collection container below the glass elbow.
12. Before proceeding to step 13, have the Chief Distillation Officer check your setup to ensure that it is correct, secure, and safe.
13. Gently turn on the faucet. Gradually increase the water flow until you have an adequate flow rate. The condenser should be completely filled with water and without bubbles. The water should be flowing strongly but not blasting.

PROCEDURE 2

Carrying Out the Distillation:

1. Record the initial temperature in the distillation head.
2. Turn on the heating mantle, and adjust the heat until the solution is boiling smoothly. Do not allow **bumping** (violent eruptions of large bubbles).
3. Monitor the temperature in the distillation head. Record the temperature of the gases that are produced at 1-minute intervals throughout the distillation.
4. Follow the guidelines below for the distillation of your specific mixture:
   - If you are distilling polluted saltwater, stop when about half the original amount remains in the still pot.
   - If you are distilling wine, stop when the temperature of the gases in the head reaches about 95°C.
   - If you are distilling crude oil, use one beaker to collect the first substances that distill through. This is called the **first cut**, or the **gasoline fraction**. When the temperature reaches about 175°C, replace the original beaker with a new beaker. This will allow you to collect the **second cut**, or the **kerosene fraction.**[1]
5. When your distillation has reached an end, turn off the heating mantle but allow the water to continue to flow.
6. When distillate (the substance collected at the end of the distillation) is no longer dripping out, remove the collection container so that the distillate can undergo further analysis.
7. As instructed by the Chief Distillation Officer, repeat this process with a second substance to be distilled or share your results with other groups of UniChem International distillation experts.

Refer to Procedure 2 section of your Report Sheet on p. 67.

PROCEDURE 3

Testing Products of Crude Oil Distillation:

1. In the fume hood, place a small amount of the first cut (the gasoline fraction) on a watch glass. Use a match to light the liquid. Describe the appearance of the liquid, how easy it was to light, and how it burned.
2. In the fume hood, place a small amount of the second cut (the kerosene fraction) on a watch glass. Use a match to light the liquid. Describe the appearance of the liquid, how easy it was to light, and how it burned.

Testing Products of Polluted Saltwater Distillation

1. In a small test tube, add 5 drops of 1.0 M $AgNO_3$ to 10 drops of filtered saltwater. Observe the formation of AgCl precipitate, indicating the presence of Cl^- ion in the saltwater.

---

[1] Note: You probably will not be able to reach high enough temperatures to produce further fractions, as is done in an industrial crude oil fractionation.

2. In a different test tube, add 5 drops of 1.0 M $AgNO_3$ to 10 drops of the distillate. Record your observations. Does this test indicate the presence of salt in the distillate?

Testing Products of Wine Distillation:

1. Look on the original bottle for the wine you distilled to find the claimed alcohol content of the wine. Record the value, with units.
2. As instructed by the Chief Distillation Officer, use a hydrometer (or 10- or 25-mL graduated cylinder and analytical balance) to measure the density of a sample of wine that has not been distilled. Record the measured wine density, with units.
3. As instructed by the Chief Distillation Officer, use the conversion table below to determine the alcohol content of the wine. Record the estimated alcohol content.
4. Repeat the steps 2–3 procedure to determine the alcohol content of the wine distillate. Record the estimated value.

Table 7.1 Conversion Table

Specific Gravity	% Alcohol by Weight	% Alcohol by Volume
0.9936	2.5	3.13
0.9897	4.8	6.00
0.9866	6.8	8.47
0.9801	11.3	14.0
0.9767	13.8	17.0
0.9734	16.4	20.2
0.9719	17.5	21.5
0.9702	18.8	23.1
0.9682	20.3	24.8
0.9658	22.1	27.0
0.9628	24.2	29.5
0.9591	26.7	32.4
0.9540	29.9	36.1
0.9472	33.8	40.5
0.9372	39.0	46.3
0.9219	46.3	53.8
0.9001	56.1	63.6
0.8631	71.9	78.2

The following table gives conversions between specific gravity and amount of alcohol (data from *Handbook of Chemistry*, at 20°C)

NAME (print) ______________________________ DATE ____________
LAST FIRST

LABORATORY SECTION ______ PARTNER(S) ______________________________

REPORT SHEET

PRE-LAB ASSIGNMENT

Pre-Lab Questions

1. Which is more volatile, water or ethanol? Explain your reasoning.

2. If a water and ethanol mixture is distilled, which substance will be collected first in the distillate? Explain your reasoning.

3. Order the following from lowest boiling point to highest boiling point:
   decane pentane propane methane octane

4. If a mixture of hydrocarbons is distilled, which substances will be collected first? Which will be collected last? Explain your reasoning.

5. Order the following from lowest boiling point to highest boiling point:
   H_2O CH_3CH_2OH NaCl $C_{10}H_{22}$

NAME (print) ______________________________ DATE ______________
LAST FIRST

LABORATORY SECTION ______ PARTNER(S) ______________________________

PROCEDURE 2

Distillation Process:

1. Record your observations for the wine distillation.

2. Record your observations for the polluted salt water distillation.

3. Record your observations for crude oil distillation.

4. Record the data for your distillation. Type of distillation: ______________

Time (min)	0.0	1.0	2.0	3.0	4.0	5.0	6.0	7.0
Temperature (°C)								

Time (min)	8.0	9.0	10.0	11.0	12.0	13.0	14.0	15.0
Temperature (°C)								

Time (min)	16.0	17.0	18.0	19.0	20.0	21.0	22.0	23.0
Temperature (°C)								

Time (min)	24.0	25.0	26.0	27.0	28.0	29.0	30.0	31.0
Temperature (°C)								

Time (min)	32.0	33.0	34.0	35.0	36.0	37.0	38.0	39.0
Temperature (°C)								

NAME (print) ____ DATE ____________________
LAST FIRST

LABORATORY SECTION ______PARTNER(S) ____

DATA ANALYSIS

1. Graph the temperature-time date for your distillation below. Be sure to label the axes, plot your data carefully, draw a smooth curve through the points, and indicate all salient features on the graph

   Type of distillation:

2. Record the results of the water purity test.

3. Record the results of the alcohol content test.

NAME (print) ____ DATE ____________________
LAST FIRST

LABORATORY SECTION ______PARTNER(S) ____

4. Describe the appearance and burning characteristics of the gasoline fraction.

5. Describe the appearance and burning characteristics of the kerosene fraction.

CONCLUSION

Explain how effective the distillations were; that is, did the polluted saltwater distillation produce pure water, did the wine distillation produce a higher-alcohol-content beverage that might be enjoyable to drink, and did the crude oil distillation separate the oil into different fractions? Based on these answers, will you recommend building the plants? Are there any other scientific observations that should be considered in making this recommendation? (Did these distillations require only the chemical substances, or were any other resources needed?) State your recommendations and the reasoning behind them in sentence format. Be sure that your conclusion is complete and concise.

Experiment 8

Coffee or Tea? Extraction of Caffeine from Coffee and Tea

PRE-LAB ASSIGNMENT

Reading:

1. Read the experiment.

Research:

2. Look up the structure of caffeine, its molecular weight, and melting point.
3. Look up the density of methylene chloride (sometimes called **dichloromethane**) and water.
4. When methylene chloride and water are mixed together, they form two layers because methylene chloride is not miscible in water. Based on your answers from question 3, which solvent will be the top layer and which will be the bottom layer. Briefly explain.
5. Remember to provide a complete reference for your source.

INTRODUCTION

Coffee or tea? This is the question Wannabe U. is trying to decide–which caffeinated beverage should be adopted as the "Official Morning Beverage" on campus. The campus is clearly divided because one group claims that since coffee has more caffeine than tea, it should be the official beverage. The other group claims that tea is just as effective as a "wake-up drink" even though it does not contain as much caffeine as coffee. Tea also contains two other compounds, theobromine and theophylline, that tend to behave as smooth-muscle relaxants. Therefore, tea has a relaxing effect, whereas coffee tends to stimulate.

To try to reinstate order and peace to the campus, the president of the university has intervened and hired the assistance of the Organic Division (OD) at UniChem International to determine which beverage contains more caffeine, coffee or tea. Both sides have agreed to support the findings of the OD from UniChem International and adopt that beverage as the "Official Morning Beverage" on campus.

Caffeine is a naturally occurring plant product from the alkaloid family. Alkaloids are a class of organic compounds where a nitrogen atom or atoms are a part of a five- or six-membered ring. Most alkaloids are bitter tasting and possess some biological activity. Other members of the alkaloid family include nicotine, cocaine, morphine, and quinine. Caffeine is considered to be stimulant because it makes us feel less fatigued and more alert. Because of caffeine's ability to increase alertness and decrease the feeling of being tired, it is the major ingredient in No Doz®.

In this experiment, you will extract caffeine from either coffee or tea using an organic solvent called **methylene chloride** that is insoluble in water. This is an important fact because the caffeine, as well as other components of coffee or tea, initially will be in the water that was used to prepare them. Caffeine is an organic compound and is more soluble in an organic solvent, such as methylene chloride than it is in water. Mixing water and methylene chloride together will form two layers and force the caffeine from the water layer to the organic layer. Since water and methylene chloride are immiscible, the two liquids can be separated with the use of a separatory funnel. The addition of anhydrous sodium carbonate will keep the other components of caffeine in the water layer. Therefore, the caffeine will be in the organic layer and the other components of coffee or tea will stay in the water layer. Allowing the organic solvent to evaporate will leave only the extracted caffeine. The amount of caffeine extracted from both coffee and tea will be weighed to determine which beverage contains more caffeine. To verify that caffeine actually was extracted, you will obtain the melting point of your caffeine sample and compare it with that of an authentic sample of caffeine.

LEARNING OBJECTIVES

Be able to:

- Perform an extraction of naturally occurring organic compounds from its matrix.
- Determine the amount of caffeine in both coffee and tea.
- Perform a melting-point determination as a way to determine purity.

APPARATUS

Chemicals:

- 50-mL prepared coffee extracts
- 2 tea bags
- 2.00 g anhydrous sodium carbonate
- 0.5 g anhydrous sodium sulfate
- ~50 mL methylene chloride

Equipment:

- 125-mL separatory flask with stopper
- Ring stand with ring clamp
- Filter paper to fit glass funnel
- Boiling stones or boiling sticks
- Hot plate
- Melting-point apparatus
- Melting-point capillary tubes
- Sample vial
- Roll of colored tape for label
- Stapler

- Are your safety glasses on?
- Methylene chloride is considered to be an irritant and toxic and should be measured in a fume hood.
- Evaporation of methylene chloride should take place in a well-ventilated area. Do not breathe in the vapors.
- Caffeine is toxic and may irritate skin and eyes. Avoid contact with clothing, skin and face.
- Please wear gloves when handling methylene chloride and caffeine.

Clean-up:

- All aqueous layers can go down the drain with water.
- Dispose of the sodium sulfate in the filter paper in a nonhazardous labeled solid-waste container.
- Dispose of any left over methylene chloride in a flammable labeled waste container.
- Tea bags can go in the trash.
- Wash your hands thoroughly with soap and water before you leave the laboratory.

PROCEDURE 1

Extraction of Caffeine from Coffee:

1. Extract caffeine from coffee or tea bags as instructed by your supervisor.
2. Add 50 mL of the prepared coffee extract to a 250-mL Erlenmeyer flask.
3. Add 2.00 g of anhydrous sodium carbonate.
4. Gently swirl the mixture until all the sodium carbonate has dissolved.
5. Go to Procedure 3.

Refer to Data Collection section of your Report Sheet on pp. 77–78.

PROCEDURE 2

Extraction of Caffeine from Tea Bags:

1. Weigh a medium-sized watch glass, and record the weight on your Report Sheet.
2. Gently remove the staples from two tea bags, and place the tea on the pre-weighed watch glass.
3. Re-weigh the watch glass with the tea and record this value on your Report Sheet.
4. Weigh the tea to the nearest 0.001 g if possible.
5. Determine the mass of the tea by difference.
6. Return the tea to the bags as evenly as possible, and staple the bags closed.
7. Place the tea bags in a 250-mL beaker so that they lie as flat as possible.
8. Add 25 mL of distilled water.

9. Weigh out 2.00 g of anhydrous sodium carbonate and add it to the beaker containing the tea bags and distilled water.
10. Gently heat the contents of the beaker on a hot plate until a gently boil is obtained.
11. Cover the beaker with a watch glass to prevent the loss of liquid owing to evaporation.
12. Continue to heat for 20 minutes.
13. Try to keep the tea bags submerged during the heating process by holding them down with a glass stirring rod.
14. At the end of the heating process, decant the liquid into a clean 50-mL Erlenmeyer flask.
15. Add an additional 10 mL of hot water to the tea bags and press the tea bags to the sides of the beaker with a stirring rod.
16. Add this tea extract to the tea extract in the Erlenmeyer flask, and allow the solution to cool to room temperature.
17. Go to Procedure 3.

Refer to Data Collection section on your Report Sheet on pp. 77–78.

PROCEDURE 3

The procedure is the same from this point on for the extraction of caffeine from either coffee or tea bags.

1. Transfer the liquid to a 125-mL separatory funnel.
2. **BE SURE THAT THE STOPCOCK IS CLOSED BEFORE ADDING THE SOLUTION.**
3. The stopcock is closed when it is parallel to the benchtop.
4. Place the separatory funnel into the ring clamp, as shown in Figure 8.1.
5. Carefully add 25 mL of methylene chloride to the separatory funnel and add the stopper.
6. Remove the separatory funnel from the ring clamp and invert the funnel. Be sure to hold the stopper in place so that no liquid leaks out.
7. Release the pressure that builds up inside the separatory funnel by slowing opening the stopcock. The opening of the stopcock should be pointed toward the back of the fume hood. *Never* point the stopcock toward yourself or your partner.
8. Close the stopcock, and gently mix the contents by inverting the separatory funnel two or three times. Be sure to hold the stopper firmly in place to avoid loss of liquid.
9. Vent the separatory funnel frequently.
10. Repeat the mixing two or three more times, venting often.
11. DO NOT MIX TOO VIGOROUSLY BECAUSE AN EMULSION MAY FORM. If an emulsion forms, please ask your supervisor for help.

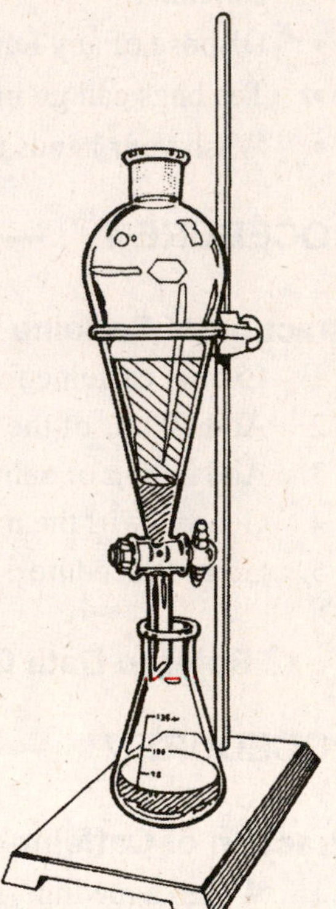

Figure 8.1 Extraction Set-up

An **emulsion** is a milky white suspension between two immiscible layers that does not separate into layers very easily.

12. Place the separatory funnel into the ring clamp, and allow the layers to separate.
13. If a small emulsion is present between the two layers, allow it to go with the bottom layer, or the methylene chloride layer.
14. Place a 125-mL Erlenmeyer flask under the opening of the stopcock.
15. Remove the stopper from the separatory funnel and carefully drain the lower organic layer into the Erlenmeyer flask. Be careful not to let any of the upper aqueous layer drain out as well.
16. Add 5 mL of methylene chloride to the water remaining in the separatory funnel and repeat the extraction procedure.
17. Combine the organic layers into the same 125-mL Erlenmeyer flask.
18. Repeat the extraction procedure a third time.
19. Add 0.5 g of anhydrous sodium sulfate to the Erlenmeyer flask containing the methylene chloride–caffeine solution.
    - ➢ The purpose of anhydrous sodium sulfate is to remove any water that may be present.

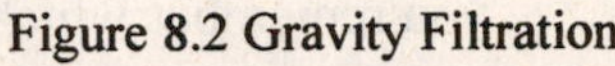

Figure 8.2 Gravity Filtration

24. Gently swirl the flask.
25. If all the anhydrous sodium sulfate clumps together or sticks to the sides of the flask add more until some of the sodium sulfate is free flowing while swirling.
26. Assemble the gravity filtration apparatus as shown in Figure 8.2.
27. Weigh a 125-mL Erlenmeyer flask that contains two boiling stones and record this weight on your Report Sheet.
28. Carefully decant the methylene chloride–sodium sulfate mixture into the funnel containing the fluted filter paper.
29. Fluted filter paper can be prepared by folding a piece of round filter paper in half several times. (See Figure 8.3) This is known as a **gravity filtration**.
30. Rinse the solid in the filter paper with 2–3 mL of methylene chloride.
31. Evaporate the methylene chloride on a hot plate set on low setting. This evaporation process **MUST TAKE PLACE IN THE FUME HOOD.** Do not heat too quickly or too hot because foaming may occur.
    - ➢ The solid that remains is crude caffeine, (not pure).

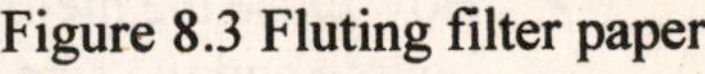

Figure 8.3 Fluting filter paper

32. Using tongs, remove the flask from the hot plate, and allow the solution to cool to room temperature.
    - ➢ Remember hot glass looks just like cold glass.
33. Re-weigh the flask and caffeine and record the weight on your Report Sheet.
34. Determine the weight of caffeine extracted from either coffee or tea bags.

Refer to Data Collection section of your Report Sheet on pp. 77–78.

PROCEDURE 4

Determining the Melting Point of Caffeine:

1. Follow the directions given to you by your supervisor for your particular melting-point apparatus.
2. Obtain an authentic sample of caffeine.
3. Place a small amount of pure caffeine and your caffeine sample on a watch glass.
4. Be careful that the samples do not mix.
5. Obtain two capillary tubes, and gently pack the capillary tubes by taking the open end and tapping into the caffeine samples.
6. Gently tap the closed end of the capillary tube against the bench op to force the caffeine sample to the bottom of the tube.
7. The capillary tubes may be dropped (bottom-end down) through a piece of glass tube that is resting on the benchtop or floor to aid in the packing process.
8. Place the packed capillary tubes into the melting point apparatus.
   - ➢ Most melting-point apparatus can hold three to five capillaries at a time. Be sure that you remember which is which.
9. Turning on the instrument will light up the samples so that the solid material inside the capillary tubes is visible when looking through the magnifying eyeglass.
10. Set the dial on the apparatus to 5 or 6.
    - ➢ This will allow the apparatus to heat up quickly but will not cause the caffeine samples to melt.
    - ➢ The melting point of the authentic caffeine sample is ~238°C. The melting point of your crude caffeine is expected to be lower than that of the pure sample.
11. You will be able to see the temperature on the thermometer and the solid in the capillary tubes simply by raising your eyes up and down.
12. Allow the temperature to increase at a quick pace until the reading on the thermometer is close to 175–185°C.
    - ➢ The correct rate of heating while taking a melting point is 1–2°C per minute once you are close to the literature value of the sample. Adjust the dial on the melting-point apparatus to a lower setting to achieve this rate.
    - ➢ All melting points are recorded in a range.
    - ➢ The range of the melting point starts at the first sign of melting and ends when the sample is melted completely.
    - ➢ How close the melting point of the sample under study is to the literature value of the sample and the size of the range is an indication of how pure the sample is.
    - ➢ A melting point that is far away from the literature value and has a large space in the range of the melting points is an indication that the sample is not pure.
    - ➢ A melting point that is close to the literature value with a small range is an indication of a pure compound.
13. Record the temperature on your Report Sheet of each sample at the first sign of melting and the temperature when the entire sample has melted.
14. Place your caffeine sample in a labeled sample vial, and give it to your supervisor.

Refer to Procedure 4 section of your Report Sheet on p. 78.

NAME (print) ______________________________ DATE ______________
LAST FIRST

LABORATORY SECTION ______ PARTNER(S) ______________________________

REPORT SHEET

PRE-LAB ASSIGNMENT

Research

1. Look up the structure of caffeine, its molecular weight, and melting point.

2. Look up the density of methylene chloride (sometimes called **dichloromethane**) and water.

3. When methylene chloride and water are mixed together, they form two layers because methylene chloride is not miscible in water. Based on your answers from question 2, which solvent will be the top layer and which will be the bottom layer. Briefly explain.

4. Remember to provide a complete reference for your source.

PROCEDURES 1–3

Data:

Caffeine was extracted from (circle one) coffee extract tea bags

Tea Bag Data:

Mass of watch glass and tea: __________g

Mass of watch glass: __________g

Mass of tea from two tea bags: __________g

NAME (print) ______________________________ DATE ______________
LAST FIRST

LABORATORY SECTION ______ PARTNER(S) ______________________________

Mass of 125-mL Erlenmeyer flask, boiling stones, and crude caffeine: __________g

Mass of 125-mL Erlenmeyer flask and boiling stones: __________g

Mass of crude caffeine extracted: __________g

PROCEDURE 4

Data Table I: Melting points of crude and pure caffeine.

Sample	Start of Melting-Point Range	End of Melting-Point Range
Crude caffeine		
Authentic caffeine		

DATA ANALYSIS

1. Which extract, coffee or tea, contained the most caffeine?

2. What will be your recommendation to the president of Wannabe U. as to the "Official Morning Beverage" on campus?

SOURCES OF ERROR

Once the data have been collected, the possibility of error in the results must be addressed. Please cite at least two random sources of error and briefly explain how each source of error may have affected your data. For a review of error analysis, see *Error Analysis in the Chemistry Laboratory* in the Appendix at the end of this manual.

NAME (print) ______________________________________ DATE ______________
LAST FIRST

LABORATORY SECTION ______ PARTNER(S) ______________________________

POST-LAB QUESTIONS

1. Why was it necessary to remove the stopper from the separatory funnel before draining the liquid?

2. Why does your caffeine look different from the authentic sample of caffeine?

3. Why was it important to add sodium carbonate to both the coffee extract and the tea extract?

4. Decaffeinated coffee and tea are available in most grocery stores. Based on the knowledge you have gained from this experiment; suggest a method for the decaffeination process.

NAME (print) ____ DATE ______________________
LAST FIRST

LABORATORY SECTION ______ PARTNER(S) ____

CONCLUSION

Based on your analysis and those of your fellow organic chemist, which item contained more caffeine, coffee or tea? How effective was the extraction process? Would the extraction process work better on a larger scale? What other things can you conclude or state? State your recommendations and the reasoning behind them in complete sentences. Be sure that your conclusion is complete and concise.

Experiment 9

Chromatography and Spectroscopy of Plant Pigments

PRE-LAB ASSIGNMENT

Reading:

1. Read the introduction to this experiment.
2. Review the concepts of polar and nonpolar substances, as well as which types of solutes dissolve in which types of solvents (see Chapter 4 of *Adventures in Chemistry*).

Research:

3. Research why objects appear a particular color, in terms of absorbed and reflected light (Chapter 17 of *Adventures in Chemistry* has some pertinent information).
4. Bring copy of a color wheel to lab.

INTRODUCTION

Chromatography, a method used to separate and identify substances, is one of the most important analytical procedures in chemistry. The Russian botanist M. S. Tsvet first used this procedure with plant pigments in 1906 and coined the term (from the Greek words for color and to write.) Chromatography typically employs a moving phase that flows past a stationary phase. Substances then separate depending on their relative attractions to either the stationary or mobile phase. Those which are more strongly attracted to the mobile phase move further in a given time, whereas those which are more strongly attracted to the stationary phase move less.

In liquid chromatography, a solvent (the mobile phase) flows through a column of a solid material (the stationary phase). During this process, each substance to be separated establishes an equilibrium between the solvent and the column material. If the equilibrium favors the solvent, then the substance flows downward with the mobile phase. If the equilibrium favors the solid, then the substance sticks to the column and does not travel downward. The equilibrium position is mostly determined by the relative strengths of attraction between substance and the solvent compared with those between the substance and the solid. In some cases, molecular weight or size may have a significant effect as well. In liquid chromatography, different solvents are used often, with the more polar substance carried through the column by the more polar solvent, and the less polar substances carried by the less polar solvent.

MacroHealth Nutritionals is developing several probiotic health food supplements containing a variety of plant extracts (from wheat grass, barley grass, spirulina etc.). MacroHealth has hired UniChem International to separate and identify the specific compounds in these plant extracts.

As a chromatography and spectroscopy expert in the Analytical Chemistry Division (ADC) of UniChem International, you have been assigned two tasks. First, you will use a small-scale chromatography column to separate the chlorophyll and carotene-xanthophyll from plant material and assess how well this technique works before doing it on a larger scale. Second, you will determine the visible spectrum of each of these substances, so that the appropriate wavelengths can be chosen to use when measuring absorbance to determine the amount of each substance.

LEARNING OBJECTIVES

Be able to:

- Create a chromatography column and use it to separate substances.
- Explain separation of substances on the basis of differing polarities.
- Determine the visible spectra of different substances.
- Explain the observed color of a substance by analyzing its visible spectrum.

APPARATUS

Chemicals:

- Grass or other green plant material
- 4:1 mixture of ligroine-acetone
- Pure acetone
- Glass wool
- Fine sand
- Silica gel

Equipment:

- Mortar and pestle
- Short-stem Pasteur pipettes
- Piece of wire
- Scoopulas
- Plastic straws
- 96-well tray
- Hole punch
- Plastic pipettes
- Spectrophotometer
- Cuvettes

- Are you wearing your safety glasses?
- Work with the solvents in a well-ventilated room.
- Don't breathe silica dust.

Clean-up:

- Dispose of all excess reagents and waste as directed by your supervisor.
- Wash your hands thoroughly with soap and water before you leave the laboratory.

PROCEDURE 1

Extracting Plant Pigments:

1. Obtain some green grass or other green plant material.
2. Place the plant material in a mortar and add about 20 mL of a 4:1 ligroine-acetone mixture. Use the pestle to grind the plant material until the liquid becomes dark green.
3. Use a plastic pipette to carefully suck up the green solution and transfer it to a small test tube. Be sure to transfer only liquid and leave behind all solid material.
4. Continue this process until the test tube is at least half full. If necessary, add more liquid and/or plant material.

Refer to Procedure 1 section of your Report Sheet on p. 86.

PROCEDURE 2

Constructing a Chromatography Column:

1. Obtain a short-stem Pasteur pipette. Place a small wad of glass wool (less than the size of a small pea) in the bottom of the Pasteur pipette. Use the piece of wire to carefully pack the glass wool down. Do not pack too tightly, because liquid must be able to flow through it.
2. Use a small scoopula to carefully transfer sand into the Pasteur pipette. Make an approximately 0.5-cm layer of sand above the glass wool, as indicated in Figure 9.1. Be sure that the layer is horizontal when the Pasteur pipette is held upright.
3. Use a (different) small scoopula to carefully transfer silica gel into the Pasteur pipette. Make an approximately 5-cm layer of silica gel above the sand, as indicated in Figure 9.1. Be sure to leave about 3 cm unfilled at the top of the Pasteur pipette. Be sure that the silica layer is horizontal.

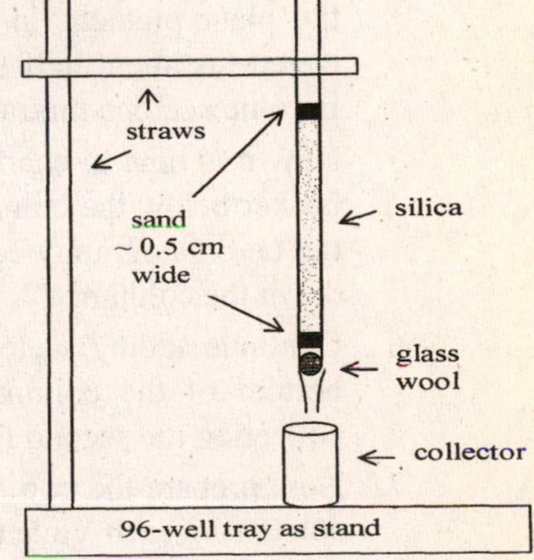

Figure 9.1 Chromatography Column

4. Once again, use the sand scoopula to create a level 0.5-cm layer of sand above the silica gel, as indicated in Figure 9.1.
5. Obtain a 96-well tray, two straws and a hole punch.
6. Cut a 10-cm piece from one of the straws, and punch a hole about 2 cm from one end. This will be the crossbar to hold your chromatography column.
7. Punch a hole through the second straw about 20 cm from one end. This will be the vertical support to hold the crossbar and column.
8. Place the end of the vertical support straw firmly in one of the holes in the 96-well tray. Insert the crossbar straw through the punch hole in the vertical straw. Then carefully insert the chromatography column through the punch hole in the crossbar straw. Adjust the supports and Pasteur pipette until the chromatography column is vertical and firmly supported. Again refer to Figure 9.1. Voilà—you have created a chromatography column!

Refer to Procedure 2 section of your Report Sheet on p. 86.

PROCEDURE 3

Separating the Plant Pigments:

1. Place a small waste beaker below the chromatography column. Use a plastic pipette to gradually add ligroine-acetone to the chromatography column. Be careful to not create bubbles or to overfill the column. When the ligroine-acetone reaches the bottom of the column, note the drip rate. About 1 drop per second is good; check with your instructor if it is not reasonably close to this.
2. Refill the column with the ligroine-acetone mixture. When the liquid level falls to near the top layer of sand, use a plastic pipette to begin adding plant pigment solution to the top of the column.
3. Continue adding the plant pigment solution to the column, making sure to always keep the liquid level above the top layer of sand.
4. The plant pigments should begin to separate. One pigment will remain virtually stationary, whereas another pigment will move down the column. When the moving pigment is near the bottom of the column, remove the waste beaker and begin collecting the plant pigment in a clean cuvette. Continue collecting this first fraction until the cuvette is about half full. (If you run out of plant pigment extract, return to adding the ligroine:acetone mixture to the column until there is enough liquid in the cuvette.)
5. Now it is time to change solvents in order to separate the other pigment. Place the waste beaker below the column, and use a clean plastic pipette to begin adding pure acetone to the chromatography column. The previously stationary pigment then will begin to move down the column.
6. Continue adding acetone to the top of the column. When the second pigment is near the bottom of the column, replace the waste beaker with a clean cuvette and continue collecting the second fraction until the cuvette is about half full.
7. Now prepare the two samples for analysis in Procedure 4. To do so, add the appropriate solvent to each cuvette (ligroine:acetone to the first fraction, and pure acetone to the second fraction) until each cuvette is at least three-quarters full (which is a sufficient amount of colored solution to be measured in the spectrophotometer).

Refer to Procedure 3 section of your Report Sheet on p. 86.

PROCEDURE 4

Determining the Absorbance Spectrum of the Plant Pigments:

1. Turn on the spectrophotometer and allow it to warm up. (The amount of time needed depends on the spectrophotometer.)
2. Add ligroine-acetone to a clean, dry cuvette until the cuvette is about three-quarters full. This is the ligroine-acetone "blank".
3. Add pure acetone to a clean, dry cuvette until the cuvette is about three-quarters full. This is the pure acetone "blank".
4. Use a Kimwipe to wipe off the outside of all four cuvettes (removing fingerprints, etc.). From now on, only hold the cuvettes by the top, as directed.
5. Adjust the wavelength on the spectrophotometer to 400 nm.
6. Place the ligroine-acetone blank in the cuvette holder of the spectrophotometer, and calibrate as directed. (The exact procedure will depend on the specific spectrophotometer.)
7. Remove the blank, and replace it with the cuvette containing the first plant pigment fraction. Record the absorbance reading. (*Note:* If the absorbance reading is above 1.5, you will first need to dilute the solution. If so, go to steps 8 and 9.)
8. If you need to dilute the plant pigment solution, pour out about half the solution (into a waste beaker). Refill the cuvette with ligroine-acetone mixture, and mix thoroughly.
9. Wipe off the cuvette, place it back in the spectrophotometer, and record the absorbance reading. If still above 1.5, repeat steps 8 and 9 until you get an appropriate reading.
10. Repeat steps 5–7 at 425, 450, 475, and up to 800 nm. Record the absorbance of the first plant pigment at each of these wavelengths. (Yes, you must continue to use the ligroine-acetone blank and recalibrate at each wavelength. However, be sure that you do not dilute the solution any further after making the 400-nm measurement.)
11. Repeat steps 5–10 using the pure acetone blank for calibration and the second plant pigment fraction to measure the absorbance.

Refer to Procedure 4 section of your Report Sheet on p. 86.

NAME (print) ______________________________ DATE ______________
LAST FIRST

LABORATORY SECTION ______ PARTNER(S) ______________________________

REPORT SHEET

PROCEDURE 1

Record your observations from the plant pigment extraction.

PROCEDURE 2

How good a job did you do on your chromatography column? Give yourself a rating (from 1 to10) for solidity of construction, drip rate, aesthetics, etc.

PROCEDURE 3

Record your observations from the plant pigment separation via chromatography.

PROCEDURE 4

Record your absorbance data.

Wavelength (nm)	First Fraction Absorbance	Second Fraction Absorbance
400		
425		
450		
475		
500		
525		
550		
575		

Wavelength (nm)	First Fraction Absorbance	Second Fraction Absorbance
625		
650		
675		
700		
725		
750		
775		
800		

NAME (print) ______________________________ DATE ______________
LAST FIRST

LABORATORY SECTION ______ PARTNER(S) ______________________________

DATA ANALYSIS

Graph the data from this experiment (wavelength on the x-axis and absorbance on the y-axis). Plot the points for the first fraction, and draw a smooth curve through the points. Using a different color, plot the points for the second fraction, and draw a smooth curve through these points. Label each curve, and label the axes, including units.

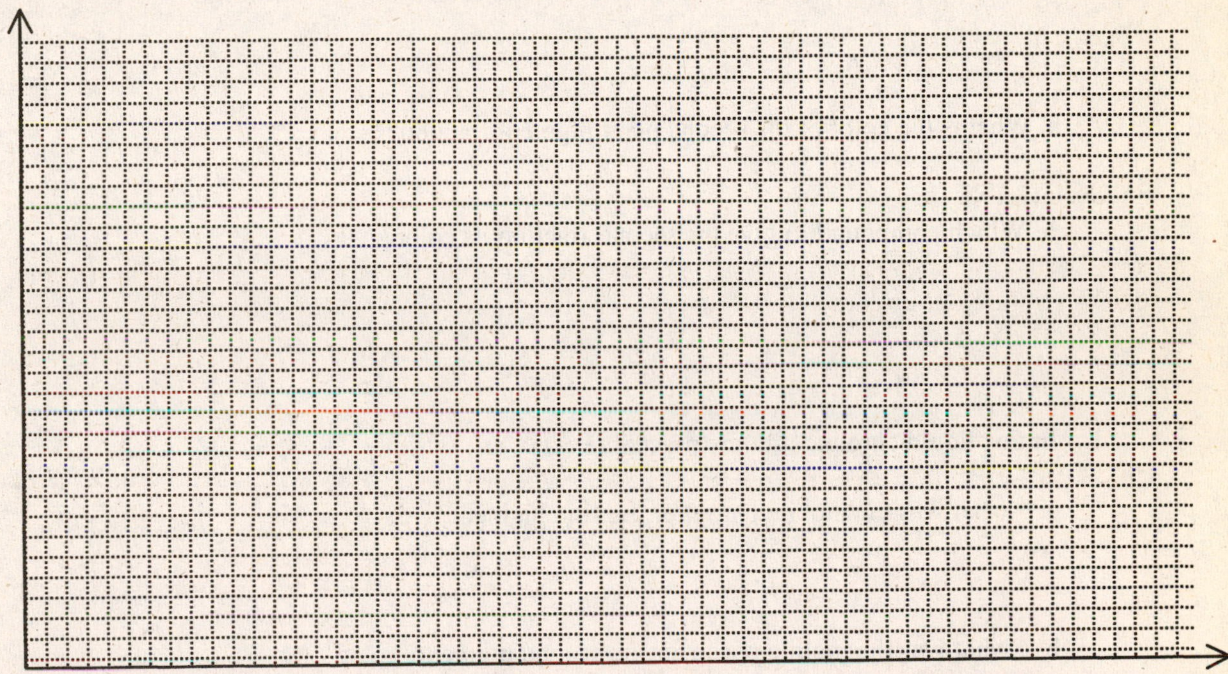

SOURCES OF ERROR

Once the data have been collected, the possibility of error in the results must be addressed. Please cite at least two random sources of error and briefly explain how each source of error may have affected your data. For a review of error analysis, see *Error Analysis in the Chemistry Laboratory* in the Appendix at the end of this manual.

NAME (print) ____________________________________ DATE ____________________
LAST FIRST

LABORATORY SECTION ______ PARTNER(S) ____________________________________

POST-LAB QUESTIONS

1. a. What is the color of the first plant pigment fraction?

   b. Which compound(s) give the first fraction its color?

2. a. What is the color of the second plant pigment fraction?

   b. Which compound(s) give the second fraction its color?

3. In this liquid chromatography procedure …

   a. what was the mobile phase for the first pigment?

   b. what was the mobile phase for the second pigment?

   c. what was the stationary phase (for both pigments)?

4. The equilibria involved during the liquid chromatography process can be expressed as

   Pigment molecules (stationary phase) $\rightleftharpoons$ pigment molecules (mobile phase)

   a. For the green pigment, which side of this equilibrium is favored when the ligroine-acetone mixture is added?

   b. For the green pigment, which side of this equilibrium is favored when the acetone is added?

NAME (print) ____ DATE ____________________
LAST FIRST

LABORATORY SECTION ______ PARTNER(S) ____

5. Which plant pigment molecules are more polar—the chlorophylls, or the carotenes and xanthophylls? Explain your reasoning. (*Note:* Acetone is a polar solvent, and ligroine is nonpolar.)

6. a. What is the predominant color of light absorbed by the first fraction ?

   b. How does this relate to the yellow color of the fraction (i.e., why does this fraction appear yellow)?

7. a. What is the predominant color of light absorbed by the second fraction?

   b. How does this relate to the green color of the fraction (i.e., why does this fraction appear green)?

NAME (print) ____ DATE ____________________
LAST FIRST

LABORATORY SECTION ______ PARTNER(S) ____

CONCLUSION

How well does your small-scale chromatography column work to separate plant pigments? Would you recommend using a larger-scale version of this process to separate larger quantities? What wavelength should be used in absorbance measurements to determine the amount of chlorophyll? What wavelength should be used for carotenes and xanthophylls? State your recommendations and the reasoning behind them in complete sentences. Be sure that your conclusion is complete and concise.

Experiment 10

Isolation of Strawberry DNA

PRE-LAB ASSIGNMENT

Reading:

1. Read the experiment.
2. Refresh your memory concerning the polarity of molecules in Chapter 4 of *Adventures in Chemistry*.

Research:

3. Look up the definition of osmosis.
4. What does the acronym DNA mean?
5. Be sure to completely reference your source. A good reference is Chapter 8 (especially Figure 8.38) of *Adventures in Chemistry*.

Acknowledgment: *Special thanks to Dr. Julie T. Millard, Lisa M. Miller, and Trevor Hanly, all of Colby College, for their work on the DNA isolation procedure.*

INTRODUCTION

The DNA division of UniChem International has been requested by the US. Department of Agriculture (USDA) to assist them in building a case against the local farmer's market that is thought to be selling genetically modified (GM) strawberries imported from Mexico. **Genetically modified foods** and **genetically modified organisms** (GMOs) refer to plants created for consumption by humans or animals that have been modified in the laboratory. These modifications are used to enhance traits of the plants, such as to improve the nutritional content, increase resistance to herbicides, and increase resistance to plant diseases caused by insects or viruses. Improving crop production was the initial objective for developing plants using GMO technology where selected individual genes are transferred from one organism into a different organism. For example, a geneticist can isolate the gene that keeps fish from freezing in very cold waters and insert that gene into the DNA of strawberries so that they can survive an early morning frost.

While over 40 genetically modified plant varieties have met all the USDA requirements for commercialization and some are even sold in grocery stores in the United States, strawberries are not one of them. Strawberries from the local framer's market were purchased by a UniChem International agent, carefully labeled, and turned over to the DNA laboratories. Working in groups of three, we have been asked us to prepare three strawberry samples for DNA sequencing to determine if the berries sold at the farmer's market have been genetically modified. The first strawberry sample will be the control sample because these strawberries were purchased at an organic grocery store and guaranteed not to be GM. The second strawberry sample is the GMO control obtained from the laboratories at the USDA. The last strawberry sample comes from the strawberries purchased from the farmer's market and labeled as the evidence sample. We have been hired to determine if the evidence strawberry sample is actually made up of normal grocery store strawberries or strawberries that have been genetically modified and thus illegal to sell in the United States. The DNA will be isolated this week and sent to the Polymerase Chain Reaction (PCR) department for amplification. Next week, the DNA will be analyzed by electrophoresis to determine if the strawberries are a GM food. After completing your analyses of the strawberry DNA, you will be able to advise the USDA as to whether or not more legal action against the owners of the local farmer's market is warranted.

To determine whether the strawberries sold at the farmer's market are indeed genetically modified, each investigative team will isolate molecules of DNA from inside the nucleus of strawberry cells. Each member of your DNA analysis team will isolate DNA from one of the three strawberry samples. The strawberry cells will be broken open with salt and detergent. The DNA released then will be purified by adding alcohol and collecting the solid material that results. The isolation method is given in Procedure 1.

All isolated strawberry DNA will be given to your lab supervisor. Your lab supervisor will deliver your samples to the PCR division of the forensic laboratory located within UniChem International. The PCR division will amplify each sample with four primer sets to produce different-sized products depending on the DNA sequence of the individual strawberries. Because some of the genes in the GM strawberries come from some other living organism, the sizes of the products will differ from an ordinary strawberry. The sizes of the products can be determined through analysis via agarose gel electrophoresis. The next time we meet, your isolated strawberry DNA will be ready for you to prepare in a gel.

LEARNING OBJECTIVES

Be able to:

- Extract DNA from strawberry cells
- Load and run an agarose gel electrophoresis
- Analyze the gel to determine if the berries are a GM food
- Report findings to the USDA

APPARATUS

Chemicals:

- Strawberry samples (labeled as either "organic grocery store control," "genetically modified strawberry control," or "farmer's market strawberry")
- Table salt
- Shampoo
- 95% ethanol or isopropanol (cold)
- Agarose gel (1 gel will be used for two groups of agents)
- 250mL running buffer (will be explained by your supervisor)

Equipment:

- Mortar and pestle
- Hot plate
- 2 medium-sized plastic funnel
- 2 50-mL plastic test tubes with cap
- 2 teaspoons
- Pieces of cheesecloth
- Glass Petri dish
- Several toothpicks (not colored)
- 50-μL micropipettes with tips
- 15- μL micropipettes with tips
- PCR rack
- PCR tube
- 1each red, green, and black permanent markers
- Gel Casting Tray
- Horizontal Electrophoresis Chamber (250-mL) with electrophoresis power supply
- Thermometer
- Staining box with an electric stirring table (shared by entire class)
- Ultraviolet (UV) Transilluminator (shared by entire class)
- Plastic spatula (shared by entire class)
- Roll of plastic wrap

- Are you wearing your safety glasses?
- Please wear gloves when handling the agarose gel.
- Do not look directly at the UV light source because it may cause damage to your eyes.
- Ethanol is flammable. Do not use a Bunsen burner.
- Please do not eat the strawberries.

Clean-up: 

- Dispose of the agarose gels and PCR tubes as instructed by your supervisor.
- All strawberries can be thrown away in the trash cans.
- Wash your hands thoroughly with soap and water before you leave the laboratory.

PROCEDURE 1

Week 1: Procedure for Isolation of DNA from Strawberries:

1. Each member of your DNA team will grind up one piece of one of the strawberry samples. You may use a mortar and pestle. Be sure you know which strawberry sample you used. Record the strawberry type on your Report Sheet. Record the names of the other DNA agents and their strawberry sample on your Report Sheet.

2. Get a hot-water bath started by filling a 400-mL beaker three-quarters full with water and placing it on a hot plate. Turn the heat to a medium setting, and heat to ~60°C.
3. Make sure that your strawberry mush is liquid enough to pour easily by adding an equal amount of water to it.
4. Using a funnel, pour about 25 mL of the mixture into a plastic test tube.
5. Add 1 teaspoon each of salt and shampoo to the contents of the plastic test tube.
6. Put the cap on tightly and carefully invert the tube five times to mix.
   - The salt helps to break open the cell by **osmosis**.
   - Because the cell contains a higher proportion of water than salt compared with the surroundings, water will exit the cell, causing it to burst open. This process is helped by the detergent in the shampoo, which dissolves the outside layer of the cell (the **membrane**, which is comprised of lipids) just like it washes away grease and dirt.
7. Put the tube into the beaker of water in the 60°C bath for 10 minutes.
   - High temperatures will (1) speed up the release of DNA from the nucleus and (2) partially denature DNase. DNase is an enzyme that can degrade DNA.
   - Once the cell breaks down completely, a cell "soup" containing the freed DNA remains.
8. While the cell soup is heating, put a piece of cheesecloth inside a plastic funnel.
9. Put the funnel on top of a second clean plastic test tube.
10. After the 10 minutes are up, carefully pour the cell soup into the funnel.
11. Let the liquid collect in the new plastic test tube. The solids will be trapped in the cheesecloth.
12. Carefully add an equal amount of ice-cold alcohol to your liquid in the plastic test tube.
13. Put the cap on the tube, and gently turn it upside down three times to mix. Mixing will disrupt the membranes of the cell. **Do not** mix too vigorously, because this will break the DNA strands. The DNA will look like a cloud of fine white threads.
14. Pour your mixture into a Petri plate. Use a toothpick to spool the DNA. Gently swirl the stick to wrap the DNA around it, much like cotton candy.
15. Using an automatic pipette, pipette 50 μL of isolated liquid portion of strawberry DNA into PCR tubes. Be careful not to draw any of the fine white threads into the pipette tip.
   - We will color code the sample types by marking the tops of the PCR tubes as follows: Red marks the organic grocery store strawberry, Green marks the USDA GM strawberry, and Black marks the evidence strawberry.
16. Place your PCR tubes into the PCR tube holder.
17. Be sure to record the letter and number of each of your samples in the PCR holder on your Report Sheet. For example, the green PCR tube is in B7.
18. Leave the PCR tubes with your supervisor.

Refer to Procedure 1 section of your Report Sheet on p. 96.

PROCEDURE 2

Week 2: Gel Electrophoresis:

Working with your same partners from the strawberry DNA isolation procedure, you will now prepare your agarose gel for sample loading. You will want to put on a pair of gloves for the following work.

1. Take a prepared 1.5% agarose gel on its casting tray, and gently place it into the center of the electrophoresis chamber.
2. Pour 250 mL of Running Buffer solution into the chamber to submerge the gel.
3. Acquire your three PCR tubes from the PCR tube holder.
4. Each investigator will pipette a sample into one well of an assigned group agarose gel.
5. Pipette 15 μL of the strawberry PCR product, taking great care to avoid drawing up air bubbles in the pipette tip. Slowly guide the pipette tip until it is sitting on the inside lip of a well, with the tip angled downward.
6. Release the sample into the well, keeping your thumb down on the pipette's release button as you draw the tip back out of the agarose gel.
7. When all samples are loaded in the agarose gel, sketch the eight well locations and label the contents of each well on your Report Sheet.
8. Place the cover on the chamber with the black lead (negative charge) nearest to the wells.
9. This will enable the amplified DNA to move toward the red lead (positive charge).
10. Plug the wire leads into the power source. Turn on the power.
11. Set voltage to 100 V, and let the electrophoresis run for approximately 1 hour. You want the blue band (which is a dye) to have moved at least half way across the agarose gel.
12. Turn off the power, and then unplug the black and red leads.
13. With your freshly gloved hands, lift the cover off the chamber.
14. Lift the gel tray out of the chamber, and then gently push the gel off its tray into a staining box.
15. Let the gel soak and stir for 10 minutes.
16. While waiting, cover the UV transilluminator with plastic wrap (this is to protect the viewing surface from chemical damage).
17. Remove the gel from stain box with a plastic spatula and place onto the transilluminator.
    - **SAFETY CAUTION**: Place the UV-ray protection shield on top of the transilluminator.
    - Looking at a UV light source without proper eye protection may cause permanent eye damage.
18. Turn on the power.
19. Make a sketch of the resulting stained gel on your Report Sheet.

Refer to Procedure 2 section of your Report Sheet on pp. 96–97.

NAME (print) ______________________________ DATE ______________
LAST FIRST

LABORATORY SECTION ______ PARTNER(S) ______________________________

REPORT SHEET

PRE-LAB ASSIGNMENT

Research:

1. Define osmosis.

2. What does the acronym DNA mean?

3. Provide a complete reference for your source(s).

PROCEDURE 1

Fill in the Data Table

Name of DNA Agents in Your Group	Strawberry Type

PROCEDURE 2

Using the template below, sketch the PCR sample well locations before electrophoresis.

Start End

Wells

0
0
0
0
0
0
0
0

NAME (print) ______________________________________ DATE ______________________
LAST FIRST

LABORATORY SECTION ______ PARTNER(S) ______________________________________

Using the template below sketch the PCR sample well locations after electrophoresis.

SOURCES OF ERROR

Once the data have been collected, the possibility of error in the results must be addressed. Please cite at least two random sources of error and briefly explain how each source of error may have affected your data. For a review of error analysis, see *Error Analysis in the Chemistry Laboratory* in the Appendix at the end of this manual.

NAME (print) ______________________________ DATE ______________
LAST FIRST

LABORATORY SECTION ______ PARTNER(S) ______________________________

POST-LAB QUESTIONS

1. When heating the DNA sample, why was it important not to let the temperature of the water-bath exceed 60°C? *Hint:* Think about what may happen to the double-helix structure of DNA at high temperatures.

2. To force strawberry DNA out of aqueous phase, ethanol was added. Draw the Lewis dot structure for water and ethanol. The chemical formula for ethanol is CH_3CH_2OH. Why is DNA less soluble in ethanol than in water? *Hint:* Which solvent is more polar?

3. Why does DNA move towards the positive electrode during electrophoresis?

NAME (print) ____ DATE ____________________
LAST FIRST

LABORATORY SECTION ______ PARTNER(S) ____

CONCLUSION

Based on your findings, what will you report to the USDA? Do the strawberries sold at the local farmer's market resemble grocery store strawberries or are they GM strawberries? What other things can you conclude or state? State your recommendations and the reasoning behind them in complete sentences. Be sure that your conclusion is complete and concise.

Experiment 11

Energy Content of Foods

PRE-LAB ASSIGNMENT

Reading:

1. Read the experiment.
2. Review Chapter 9 of *Adventures in Chemistry*, particularly the description of fats and carbohydrates and their caloric content.

Question:

3. Answer the Pre-Lab Questions.

INTRODUCTION

Human beings, whether somnolent college students sleeping through their chemistry lectures or Tour de France cyclists racing up the Col du Galibier, require energy to live and function. We are marvelously complex biochemical engines that use carbohydrates, proteins and fats to fuel our voluntary activities (such as walking to lab or bicycling 150 miles) and to sustain our basic metabolic processes (such as breathing, circulating blood, and maintaining body temperature).

The energy content of food is measured in nutritional Calories, with one nutritional Calorie equal to 1000 calories. (A **calorie** is the amount of energy needed to raise the temperature of 1 g of water by 1°C.) Humans use about 1 Calorie per kilogram of body weight simply to provide energy for our basic metabolic processes. Additional energy is required to fuel our other activities—for instance, walking or running uses about 100 Calories per mile, with the exact amount depending on both speed and body size. If a person's caloric intake equals his or her energy expenditure, then body weight remains constant. However, any caloric intake beyond one's energy requirements is stored as adipose tissue (body fat), with about 4000 extra Calories producing about 1 pound of fat.

The Food and Drug Administration (FDA) requires nutritional labeling for most prepared foods. Among other things, these labels provide the amounts of carbohydrates, fats and proteins in each serving, as well as the total number of Calories and the number of Calories from fat. The caloric content of food is determined via calorimetry. In a **bomb calorimeter**, food is burned in a thick metal cylinder that has been pressurized with oxygen and that is completely surrounded by water. A much simpler method involves using the energy produced by burning food to heat water under normal atmospheric conditions. Both methods require careful calibration by using a substance whose heat of combustion is already known accurately.

The Victuals Analysis Division (VAD) of UniChem International has been assigned the task of measuring the caloric content of a variety of foods. Specifically, the VAD has been asked to determine the caloric content of a number of both high- and low-fat (i.e. high carbohydrate) foods, to compare the experimentally determined values with those claimed on the food packages, and to compare the caloric contents of the high- and low-fat foods. Although food is "burned" somewhat differently in the body than in the tabletop combustion calorimetry method you and your fellow calorimetry technicians in the VAD will use, this technique can be used to determine the approximate energy content of foods.

LEARNING OBJECTIVES

Be able to:

- Construct and calibrate a simple calorimeter.
- Use combustion calorimetry to estimate the caloric content of foods.
- Distinguish between high- and low-fat foods in terms of their energy content.

APPARATUS

Chemicals:

- Various high-fat foods (e.g., peanuts, cashews, potato chips, etc.)
- Various low-fat foods (e.g., marshmallows, baked chips, popcorn, etc.)
- Cooking oil

Equipment:

- Wire food holder
- Oil burner
- Aluminum can with wire hanger
- Ringstand and clamps
- Thermometer (or temperature probe)

- Are your safety glasses on?
- Work in the fume hood or in a well-ventilated room.

Clean-up:

- Dispose of all excess reagents and waste as directed by your supervisor.
- Wash your hands thoroughly with soap and water before you leave the laboratory.

PROCEDURE 1

Determining the Efficiency of the Calorimeter:

1. Obtain a small oil burner (or construct one as directed by your supervisor).
2. Set up the calorimeter apparatus as shown in Figure 11.1. Place the oil burner on the base of the ring stand. Use a clamp to suspend the aluminum can so that its bottom is centered about 4 cm above the wick of the oil burner.
3. Use a match to light the oil burner. Adjust the can vertically so that the bottom of the can is 1 cm above the tip of the flame. Then blow out the flame and allow the burner and can to cool.
4. Once it has cooled, determine the initial weight of the oil burner to the thousandth of a gram.
5. Weigh the empty aluminum can to the thousandth of a gram.
6. Add about 50 mL of cold water (between 10 and 15°C) to the can. Weigh the can plus water to the thousandth of a gram.
7. Put a small piece of rubber tubing over the tip of a thermometer. This will protect the thermometer, which will be used for stirring as well as for measuring the temperature.
8. Determine the initial temperature of the water. Be sure that the thermometer is in the water long enough to equilibrate with the actual water temperature before recording.
9. Re-hang the aluminum can containing water, and place the thermometer in the can. Replace the oil burner directly below the can so that you once again have the setup in Figure 11.1.
10. Use a match to light the oil burner. Use the thermometer to carefully stir the water while the can and water are heating.
11. When the temperature has risen to about 30°C, blow out the oil burner flame. Keep stirring with the thermometer until the temperature stops rising. Record the highest temperature reached by the water.
12. When cool enough to handle, determine the final mass of the oil burner.
13. Dump the water out of the can. Clean any soot from the bottom of the can, and then dry the outside of the can.
14. Repeat this procedure to obtain a second determination of the oil combustion data.

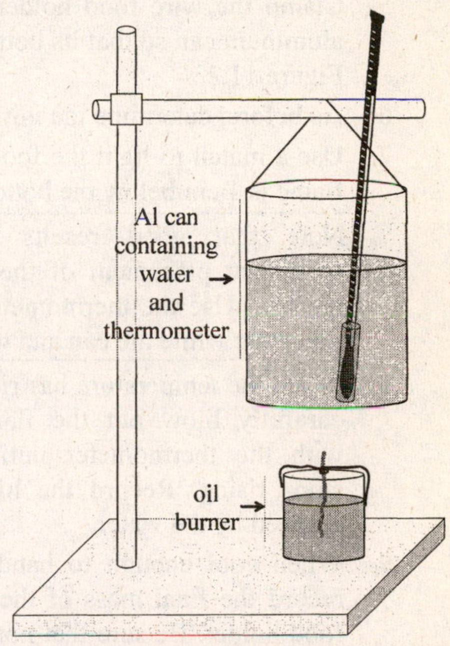

Figure 11.1
Calorimeter with oil burner

Refer to Procedure 1 section of your Report Sheet on p. 107.

PROCEDURE 2

Determining the Caloric Content of a High-Fat Food

1. Obtain a few small pieces of high-fat food.
2. Obtain a wire food holder that can securely hold a small piece of high-fat food during combustion (or construct one as directed). Place the food on the holder.
3. Determine the initial weight of the food sample plus wire holder to the thousandth of a gram.
4. Add about 50 mL of cold water (between 10 and 15°C) to the can. Weigh the can plus water to the thousandth of a gram.
5. Clamp the wire food holder about 10 cm above the base of the ring stand. Hang the aluminum can so that its bottom is about 4 cm above the food in the wire holder. Refer to Figure 11.2.
6. As before, determine the initial temperature of the water.
7. Use a match to light the food sample. Quickly adjust the apparatus so that the tip of the flame is 1 cm below the bottom of the can, as in Procedure 1.

> Note that good results will depend on consistent placement of the flame relative to the can. Use the thermometer to carefully stir the water while the can and water are heating.

8. When the temperature has risen to about 30°C, carefully blow out the flame. Keep stirring with the thermometer until the temperature stops rising. Record the highest temperature reached by the water.
9. When cool enough to handle, determine and record the final mass of the food sample plus food holder. Be sure that none of the unburned food falls off and is lost.
10. Dump the water out of the can. Clean any soot from the bottom of the can, and then dry the outside of the can.
11. Remove any unburned food from the wire holder. Clean and dry the holder.
12. Repeat this procedure to obtain a second determination of the food combustion data.

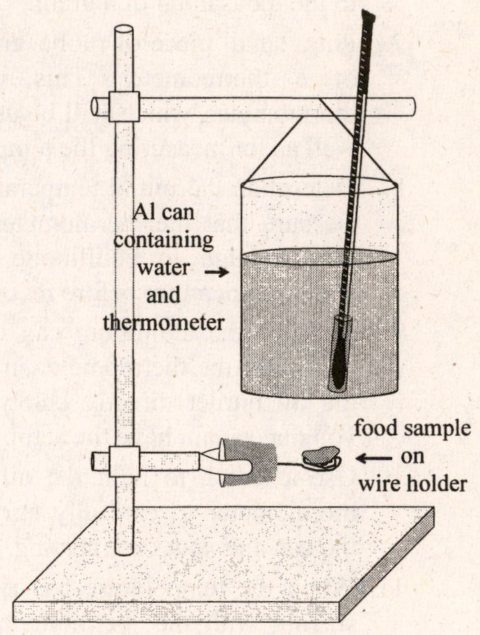

Figure 11.2
Calorimeter with food sample

Refer to Procedure 2 section of your Report Sheet on pp. 107–108.

PROCEDURE 3

Determining the Caloric Content of a Low-Fat Food

1. Obtain a few small pieces of low-fat food.
2. Repeat steps 2 – 12 of Procedure 2 to obtain combustion data for the low-fat food.

Refer to Procedure 2 section of your Report Sheet on pp. 107–108.

DATA ANALYSIS

Calculations:

1. For the first oil burner combustion trial, calculate the heat absorbed by the calorimeter:
   a. Calculate the heat absorbed by the water. Use the equation $q_{water} = (m \times c \times \Delta T)_{water}$, where q is the heat absorbed by the water (in Calories), m is the mass of water, c = 0.00100 Cal/(g • °C) is the specific heat of water, and ΔT is the change in temperature.
   b. Calculate the heat absorbed by the can. Use the equation $q_{can} = (m \times c \times \Delta T)_{can}$, where q is the heat absorbed by the aluminum can (in Calories), m is the mass of water, c = 0.00021 Cal/(g • °C) is the specific heat of water, and ΔT is the change in temperature.
   c. Add the heats absorbed by the water and the can to obtain the total heat absorbed.
2. Divide the total heat absorbed by the mass of oil burned to determine your experimental energy content of the oil in Calories per gram.
3. Repeat this process (steps 1–2) for the second oil burner combustion trial.
4. Calculate the average of the two oil combustion trials.
5. Determine the efficiency of your calorimeter by dividing your (average) experimental value for the energy content of oil (step 4 result) by the accepted value for the energy content of oil (9 Cal/g).
6. Calculate your experimental value for the energy content of the high-fat food by repeating the step 1–4 calculations, but using the high-fat food combustion data.
7. Then calculate your final value for the energy content of the high-fat food by dividing your experimental value (step 6 result) by the efficiency (step 5 result).
8. Calculate your experimental value for the energy content of the low-fat food by repeating steps 6 – 7, but using the low-fat-food combustion data.
9. As directed by your VAD supervisor, share your food energy data with your fellow calorimetry technicians. Record the results for each group in the given table.

Refer to Data Analysis section of your Report Sheet on pp. 108–110.

NAME (print) ______________________________________ DATE ______________
LAST FIRST

LABORATORY SECTION ______ PARTNER(S) ________________________________

REPORT SHEET

PRE-LAB ASSIGNMENT

Pre-Lab Questions

1. How many Calories are provided by 1 gram of each of the three macronutrients?

   1 g fat = _____ Cal 1 g carbohydrate = _____ Cal 1 g protein = ____ Cal

2. One serving of New England clam chowder contains 16 g of carbohydrate, 14 g of fat, and 6 g of protein.

   a. Calculate the number of Calories per serving provided by carbohydrates. Show your work.

   b. Calculate the number of Calories per serving provided by fats. Show your work.

   c. Calculate the number of Calories per serving provided by proteins. Show your work.

   d. Calculate the total number of Calories per serving. Show your work.

3. "Calorically dense" foods are those that provide a lot of Calories per serving. Do calorically dense foods generally have a high carbohydrate content, a high fat content, or a high protein content? Explain your reasoning.

NAME (print) ______________________________ DATE ____________
LAST FIRST

LABORATORY SECTION ______ PARTNER(S) ______________________________

Procedure 1 Data

	Trial 1	Trial 2
Weight of oil burner *before* burning:	________	________
Weight of oil burner *after* burning:	________	________
Weight of oil burned:	________	________
Weight of aluminum can plus water:	________	________
Weight of aluminum can:	________	________
Weight of water:	________	________
Final temperature (T_{final}):	________	________
Initial temperature ($T_{initial}$):	________	________

Procedure 2 Data

	High-Fat-Food	Low-Fat-Food
Type of food:	________	________
Weight of food + holder *before* burning:	________	________
Weight of food + holder *after* burning:	________	________
Weight of food that actually burned:	________	________

NAME (print) ____________________________ DATE ______________
LAST FIRST

LABORATORY SECTION ______ PARTNER(S) ______________________

Weight of aluminum can plus water:	________	________
Weight of aluminum can:	________	________
Weight of water:	________	________
Final temperature (T_{final}):	________	________
Initial temperature ($T_{initial}$):	________	________

DATA ANALYSIS

Procedure 1 Calculations

	Trial 1	Trial 2
Temperature change (ΔT):	________	________
Heat absorbed by water:	________	________

Show calculations:

Heat absorbed by can:	________	________

Show calculations:

Total heat absorbed:	________	________

NAME (print) ____ DATE ______________________
LAST FIRST

LABORATORY SECTION ______PARTNER(S) ______________________________________

Heat of combustion
per gram of oil: ____________________ ____________________

Show calculations:

Average heat of combustion
per gram of oil: ____________________

Efficiency of calorimeter: ____________________

Show calculation:

Procedure 2 Calculations

High-Fat-Food **Low-Fat-Food**

Temperature change
(ΔT): ____________________ ____________________

Heat absorbed
by water: ____________________ ____________________

Show calculations:

Heat absorbed
by can: ____________________ ____________________

Show calculations:

NAME (print) ______________________________________ DATE ____________________
LAST FIRST

LABORATORY SECTION ______PARTNER(S) ______________________________

Total heat absorbed: ____________________ ____________________

Heat of combustion per gram of food: ____________________ ____________________

Show calculations:

Record your results and the results of the other groups (in Calories per gram) in the table below.

Determine the *average* energy content for each food and record in the appropriate space.

High-Carbohydrate Foods				**High-Fat Foods**			
Food =		**Food =**		**Food =**		**Food =**	
Group	**Energy Content**	**Group**	**Energy Content**	**Group**	**Energy Content**	**Group**	**Energy Content**
Average		**Average**		**Average**		**Average**	

NAME (print) ____ DATE ______________________
LAST FIRST

LABORATORY SECTION ______ PARTNER(S) ____

SOURCES OF ERROR

Once the data have been collected, the possibility of error in the results must be addressed. Please cite at least four random sources of error and briefly explain how each source of error may have affected your data. For a review of error analysis, see *Error Analysis in the Chemistry Laboratory* in the Appendix at the end of this manual.

POST-LAB QUESTIONS

1. Which food had the highest energy content? Which had the lowest energy content?

2. a. Which of the foods in this experiment had a high-fat content?

   b. Which of the foods in this experiment had a high-carbohydrate content?

   c. From your experimental results, what generalization can you make about the relative energy content of high-fat foods and high-carbohydrate foods?

NAME (print) ______________________________________ DATE ____________________
LAST FIRST

LABORATORY SECTION ______ PARTNER(S) ______________________________________

4. Use the nutritional information given on the packages to calculate the actual energy content (in Calories per gram) for both marshmallows and peanuts. Show your work.

Marshmallows Peanuts

5. Compare these actual energy content values calculated in question 4 with the class averages for the experimental caloric content (in Calories per gram) of marshmallows and peanuts.

a. Do your results support the generalization you made in question 2.c.?

b. How closely do the experimental values agree with the actual energy values? Which are higher and which are lower? Why?

CONCLUSION

Describe what you and your fellow calorimetry experts have determined about the caloric content of the high- and low-fat foods. State your conclusions and the reasoning behind them in complete sentences. Be sure that your conclusion is complete and concise.

Experiment 12

Fruity Esters

PRE-LAB ASSIGNMENT

Reading:

1. Read the experiment.

Research:

2. Look up the structures for all the carboxylic acids and alcohols listed in Table 12.1.
3. Draw the structures in the spaces provided in Data Table I on p. 117.
4. Provide a complete reference(s) for your source(s).

Assignment:

5. Using the chemical reaction below as a guide, write a chemical reaction for the ester formed when acetic acid reacts with isoamyl alcohol in the space provided on your data sheet. See Chapter 7 of *Adventures in Chemistry* for help.

INTRODUCTION

The manufacturer of ChewaLot bubble gum wants to design a new line of bubble gum that targets 8- to 12-year-olds. A survey of this age group indicates that fruit flavor bubble gums rank higher than mint or cinnamon flavors. The manufacturer has hired the services of the Food Science Division (FSD) of UniChem International to develop four new flavors of bubble gum. A logical starting point for the new gum flavors is esters. Esters belong to a group of organic compounds that have the general formula RCO_2R', where R and R' are alkyl or aryl groups. Esters tend to possess very sweet, fruity aromas and are used commonly in the preparation of artificial food flavors that are used as flavorings in soft drinks, cake mixes, ice creams, and gum. Esters also occur in nature and can be found in fruits, flowers, and animals.

The reaction through which esters are formed is called **esterification**. We will use this very useful reaction again in Experiment 15, when aspirin or acetylsalicylic acid will be synthesized. Esterification involves the reaction of a carboxylic acid with an alcohol in the presence of an acid catalyst. Notice in the equation below that a molecule of water is produced as a by-product of this

Leaves as water

O, O, H, H—OCH_3, H^+ catalyst, O, O, CH_3, + H_2O

Benzoic Acid
Carboxylic acid

Methanol
Alcohol

Methyl benzoate
Ester

reaction. It appears that the water is formed when the elements OH from the carboxylic acid bonds with the H from the alcohol. The ester is formed when the remaining parts of the carboxylic acid and the alcohol come together.

LEARNING OBJECTIVES

Be able to:

- Draw the structures of carboxylic acids, alcohols and esters
- Write equations for the formation of esters.
- Identify esters by their characteristics aromas.

APPARATUS

Chemicals:

- Labeled dropper bottles containing concentrated acetic acid, butyric acid, and formic acid
- Labeled dropper bottles containing hexanol, propanol, isoamyl alcohol, octanol, benzyl alcohol, ethanol, and methanol
- Labeled dropper bottles containing 6 M sulfuric acid

Equipment:

- Hot plate

- Are your safety glasses on?
- When noting the aromas of the esters, fan across the top of the test tube and note the aromas. **DO NOT** smell the contents of the test tubes directly.
- Butyric acid has a strong and repugnant odor and should be dispensed in a fume hood or in a well-ventilated room.
- Concentrated acetic acid and 6 M sulfuric acid have very strong vapors that may irritate the eyes and nose. Use in a fume hood or in a well-ventilated room. Use care when handling acetic acid and sulfuric acid or wear gloves. Wash your hands after handling these acids.
- Hot glassware looks the same as cold glassware. Use test tube clamps when removing test tubes from the hot-water bath.
- Do not taste the esters. They may smell like fruits, but they are still chemicals.

Clean-up: 

- Dispose of all esters in a container labeled "Organic Waste."
- Place any excess acids in a container labeled as "Acid Waste."
- These containers can be found in the fume hood.
- Wash your hands thoroughly with soap and water before you leave the laboratory.

Table 12.1: Carboxylic Acid and Alcohol Combinations

Mixture	Carboxylic Acid	Alcohol	Fragrance
1	Acetic acid	Hexanol	Cherry
2	Acetic acid	Propanol	Pear
3	Acetic acid	Isoamyl alcohol	Banana
4	Acetic acid	Octanol	Orange
5	Acetic acid	Benzyl alcohol	Peach
6	Butyric acid	Ethanol	Pineapple
7	Butyric acid	Methanol	Apple
8	Formic acid	Methanol	Raspberry

PROCEDURE 1

Synthesis of Esters:

1. Fill a 400-mL beaker half full with water and heat the water to a boil on a hot plate.
2. Label four 13 × 100 mm test tubes according to the mixtures you will be using to prepare the esters. For example, Mixture 1 corresponds to the ester formed from acetic acid and hexanol. The esters you will prepare will be determined by your supervisor.
3. Add 2 mL of the carboxylic acid to the correspondingly labeled test tube.

   ⚠ If you spill any of the carboxylic acids, alert your supervisor immediately.
4. Add 3 mL of the corresponding alcohol from the same mixture number.
5. Gently agitate the test tube to mix the contents.
6. In the fume hood **slowly** add 15 drops of 6 *M* sulfuric acid to the test tube containing the carboxylic acid–alcohol mixture.

   ⚠ If you spill any H_2SO_4, alert your supervisor immediately.
7. Gently agitate the test tube to mix the contents.
8. Place the test tube in the hot-water bath from step 1 for ~ 15 minutes.
9. At the end of the heating time, remove the test tube from the hot water bath using a test tube clamp.
10. **Cautiously** fan the vapors toward you, as demonstrated by your supervisor.
11. If no vapor is detected, add ~15 mL of hot water from the hot-water bath to a 50-mL beaker and quickly pour the contents of the test tube into the hot water.
12. Repeat step 10.
13. Record the aroma of the ester on your Report Sheet.
14. Repeat steps 3– 11 with at least three different carboxylic acid–alcohol combinations.
15. Record the aromas of the esters on your Report Sheet.

Refer to Data Table II of your Report Sheet on p. 118.

16. Exchange data with your coworkers to obtain the aromas of the remaining four esters you did not prepare.

Refer to Data Table III of your Report Sheet on p. 118.

NAME (print) ____ DATE ____________________
LAST FIRST

LABORATORY SECTION ______ PARTNER(S) ____

REPORT SHEET

PRE-LAB ASSIGNMENT

Research:

1. Look up the structures for all the carboxylic acids and alcohols listed in Table 12.1.
2. Draw the structures and names in the spaces provided in the Data Table I below.
3. Provide a complete reference(s) for your source(s).

Data Table I: Names and Structures of Carboxylic Acids and Alcohols

Mixture Number	Name and Structure of Carboxylic Acid	Name and Structure of Alcohol
1		
2		
3		
4		
5		
6		
7		
8		

NAME (print) ____ DATE ______________________
LAST FIRST

LABORATORY SECTION ______ PARTNER(S) ____

Assignment:

1. Using the esterification reaction on page 113 of this experiment as a guide, write a chemical reaction for the ester formed when acetic acid reacts with hexanol in the space provided below.

Data Table II: Esters and Their Aromas

Ester Number	Name and Structure of Ester	Aroma of Ester

Data Table III: Other Esters and Their Aromas

Ester Number	Name and Structure of Ester	Aroma of Ester

NAME (print) ______________________________ DATE ______________________

LAST FIRST

LABORATORY SECTION ______ PARTNER(S) ______________________________

POST-LAB QUESTIONS

1. How did the aroma of your esters compare with the aromas listed in Table 12.1?

2. There are many commercially available products on the market today that use esters as flavorings. Name at least two, and indicate the flavoring used (better yet, name the ester).

3. Oil of wintergreen is an ester used in deep-heating rubs to remove soreness from muscles. It is prepared from salicylic acid and methanol, whose structures are given below. What is the structure of the ester prepared from salicylic acid and methanol?

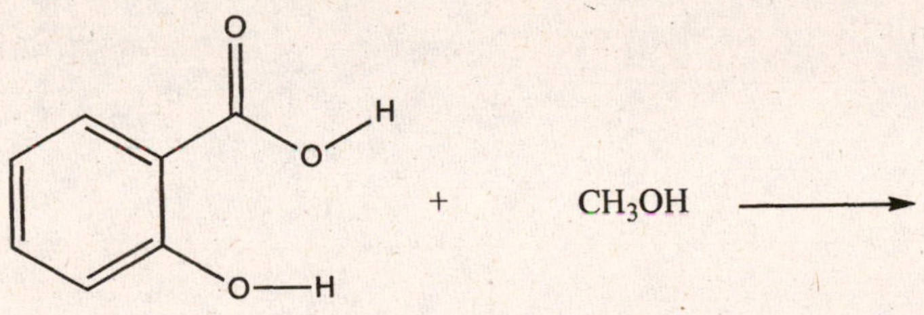

O

H

O

O—H

$+ \quad CH_3OH \longrightarrow$

NAME (print) ______________________________ DATE ______________
LAST FIRST

LABORATORY SECTION ______ PARTNER(S) ______________________________

4. Esters are named by giving the name of the alcohol first followed by the name of the carboxylic acid. The "ic acid" portion of the carboxylic acid is dropped and replaced with "ate." How would you name the ester formed in question 3?

CONCLUSION

Which esters would you recommend to the manufacturer of ChewaLot bubble gum for its new line of fruity bubble gum flavors? Briefly explain your choices. Are there any other suggestions or comments that you would like to make to the manufacturer concerning their new line of bubble gum flavors? What other things can you conclude or state? State your conclusions and the reasoning behind them in complete sentences. Be sure that your conclusion is complete and concise.

Experiment 13
Vitamin C Analysis

PRE-LAB ASSIGNMENT

Reading:

1. Read the experiment.

Assignment:

2. If instructed, obtain a lightly colored sample of fruit juice containing vitamin C.
3. Research the vitamin C content of various foods and beverages.
4. Research the antioxidant function of vitamin C.

Question:

5. Answer the Pre-Lab Questions.

INTRODUCTION

Until the end of the eighteenth century, sailors on long voyages suffered (and often died) from scurvy, a disease caused by a deficiency of vitamin C. Scurvy is characterized initially by swollen gums, weakness, fatigue, and a tendency to bruise easily and not heal well from wounds. If untreated, this condition can lead to tooth loss, anemia, muscle degeneration, convulsions, and even death.

In 1753, after years of experimenting with the diets of sailors suffering from scurvy, James Lind demonstrated that eating citrus fruit would cure scurvy. However, the identity of the actual antiscurvy component of citrus fruits remained unknown until 1928, when Albert Szent-Györgyi isolated hexuronic acid (now called **ascorbic acid** or, more commonly, vitamin C) in 1928.

Like all vitamins, vitamin C ($C_6H_8O_6$) is required in trace amounts for good health and (in humans at least) must be obtained through the diet. It is found in all citrus fruits and in most fresh vegetables. The Recommended Daily Allowance (RDA) is 60 mg of vitamin C for adults, although there is some evidence that there may be health benefits in taking daily doses larger than the RDA.

The Food and Drug Administration (FDA) is responsible for regulating and monitoring the safety and quality of a wide range of products that affect human health. Among their many functions, FDA scientists decide whether drugs such as Prevacid, Vioxx, and Viagra are effective and safe and measure the actual vitamin and mineral contents of consumer products to ensure that they contain the claimed amounts of each substance.

Owing to understaffing, the FDA recently has been forced to outsource some of its duties. Because of the excellent reputation of its Victuals Analysis Division (VAD), UniChem

International has been selected to measure the vitamin content of a variety of foodstuffs. As a newly hired quality-control chemist in the VAD of UniChem International, you have been assigned to determine the vitamin C content of a commercial beverage.

You will determine the vitamin C content by using the color changes that occur during an oxidation–reduction titration of vitamin C with the organic dye 2,6-dichlorophenol-indophenol (DIP). Solutions of vitamin C, in either its oxidized or reduced forms, are colorless. But the oxidized form of DIP has a deep-blue color in basic solution and appears red in acidic solution, whereas the reduced form of DIP is colorless in both acidic and basic solutions.

> A **titration** is a quantitative volumetric analysis method in which the volume of reagent (the **titrant**) needed to react with a particular substance to be measured (the **analyte**) is determined precisely. The **equivalence point** occurs when exactly enough titrant has been added to completely react with the analyte. In practice, this point is found by measuring an **endpoint**, which is characterized by an abrupt change in an observable property of the solution, such as its color.

When DIP is added to an acidified solution containing vitamin C, the vitamin C is oxidized to dehydroascorbic acid and DIP is changed to its reduced form—and you will observe the **BLUE** form of DIP being converted to its **COLORLESS** form. This will continue until you have added enough DIP to completely react with all the vitamin C in the solution (i.e., you have reached the equivalence point). After that, if you add any more DIP, you will observe it being converted to its **RED** acidified form (i.e., you have reached the endpoint).

The reaction, which is one of the official methods used by the FDA, is summarized below:

2,6-dichlorophenol-indophenol (oxidized form in basic solution) BLUE	→	2,6-dichlorophenol-indophenol (oxidized form in acidic solution) RED	+	Vitamin C (ascorbic acid) (reduced form)	→	2,6-dichlorophenol-indophenol (reduced form in acidic solution) COLORLESS	+	dehydro-ascorbic acid (oxidized form)

Note that vitamin C also can be oxidized by many substances other than DIP, including oxygen in the atmosphere and a number of ions, such as Fe^{2+} and Cu^{+}, that can be found in aqueous solutions.

LEARNING OBJECTIVES

Be able to:

- Describe some of the chemistry of vitamin C.
- Describe some of the effects of a deficiency of vitamin C in the diet.
- Perform an oxidation–reduction titration (quantitative volumetric analysis technique).
- Perform titration calculations to determine the amount for vitamin C in a substance.
- Compare the experimentally determined and claimed vitamin C contents.
- Determine when a dichlorophenol-indophenol–vitamin C titration would be effective.

APPARATUS

Chemicals:

- Sample of commercial fruit juice containing vitamin C (colorless, or lightly colored)
- Metaphosphoric acid-acetic acid (MAA) reagent
- 2,6-Dichlorophenolindophenol (DIP) reagent

Equipment:

- 50-mL burette
- Burette clamp and stand
- 10-mL volumetric pipette and pipette bulb
- Cheesecloth (if using juices containing pulp)
- (Optional) Stir plate and stir bar

- Are your safety glasses on?
- If you spill metaphosphoric acid-acetic acid, clean up using acid neutralizer, and alert your supervisor.
- The 2,6-dichlorophenolindophenol reagent can stain your clothing.
- You may want to wear gloves when handling this reagent.

Clean-up: 

- Dispose of all excess reagents and waste as directed by your supervisor.
- Wash your hands thoroughly with soap and water before you leave the laboratory.

PROCEDURE 1

1. Obtain a sample of fruit juice. Record the brand and type of juice used, as well as the vitamin C information from the nutritional label.

2. Use a clean, dry graduated cylinder to add 50.0 mL of fruit juice to a 250-mL Erlenmeyer flask.
   - If the juice contains pulp, strain it through several layers of cheesecloth first.
3. Use a clean, dry graduated cylinder to add 50.0 mL of MAA reagent to the 250 mL Erlenmeyer flask. Swirl the fruit juice–MAA solution until it is well mixed.
   ⚠ If you spill any MAA, alert your supervisor immediately.
4. Obtain a clean 50-mL burette, a burette clamp, and a burette stand. If instructed to do so, obtain a stir plate and stir bar as well.
5. Use a clean, dry beaker to obtain about 75 mL of DIP reagent. Record the conversion factor listed on the label of the DIP container.
6. Use about 5 mL of DIP reagent to rinse the burette, as instructed.
7. Repeat the step 6 rinsing procedure with another 5 mL of DIP.
8. Set up the filled burette, the burette clamp and the burette stand as shown in Figure 13.1. If instructed to do so, set up a stir plate as well.
9. Place a beaker under the tip of the burette and open the stopcock. Allow the DIP to flow out of the burette until all air bubbles have been flushed out and the meniscus is on the graduated scale.
   - Note: Do not try to stop the flow on an initial burette reading of 0.00 mL, because this will introduce a bias into your measurements.
10. Estimate the initial burette reading to the nearest hundredth of a milliliter (be sure to read at the bottom of the meniscus). Record this value.
11. As instructed, use a 10-mL volumetric pipette and pipette bulb to transfer 10.00 mL of the fruit juice–MAA solution (which you made in step 3) into a clean, dry 125-mL Erlenmeyer flask.
12. Place the 125-mL Erlenmeyer flask on a piece of white paper (so that you can see the colors more easily) directly underneath the burette. (If you are using a stir plate, place the Erlenmeyer flask on top of the stir plate.)
13. Open the stopcock to begin adding DIP reagent to the fruit juice solution. The fruit juice solution will turn pink initially where the DIP reagent flows into it, and then revert to its original color when the flask contents are mixed.

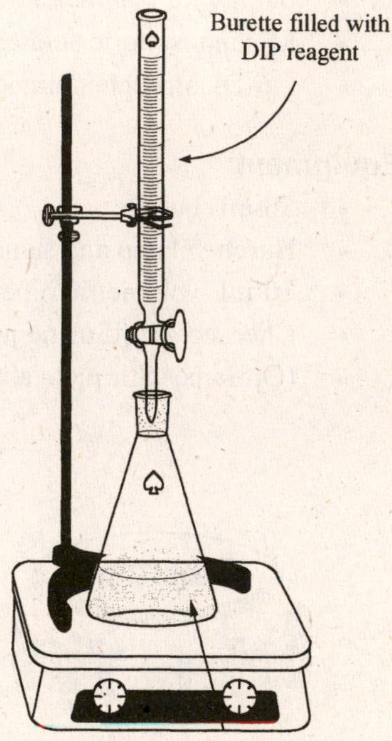

Figure 13.1
Fruit juice titration setup

14. Continue adding DIP reagent to the fruit juice solution, with constant swirling of the flask to mix the two solutions. (If you are using a stir plate, use a stir bar in the Erlenmeyer flask to provide the mixing.) Initially the pink color will disappear rapidly, and the DIP reagent can be added quickly. As you continue titrating, the increasing persistence of the pink color indicates that you are nearing the endpoint and will need to add the DIP reagent in smaller quantities.
15. When a faint but distinct pink color forms throughout the entire solution and lasts for at least 5 seconds, you have reached the endpoint. Record the burette volume at the

endpoint. (Again, be sure to use the bottom of the meniscus to estimate the volume to the hundredth of a milliliter.)

16. Calculate the actual volume of DIP reagent used in the titration, and record.
17. Refill the burette, and repeat this procedure (steps 11–16) with a second fruit juice–MAA sample. If your second DIP reagent volume is within 0.5 mL of the first trial volume, you may proceed to the calculations and questions. If not, you may need to do a third trial—please check with your supervisor.

Refer to Titration Data section of your Report Sheet on p. 127.

DATA ANALYSIS

Calculations:

Use your experimental data to determine the amount of vitamin C in one serving of your fruit juice. Be sure to show your values with units and to the correct number of significant figures, and show the setup of all calculations.

1. Use the results from the two (or three) trials to calculate the average volume of DIP reagent required to titrate the vitamin C in the fruit juice samples.
2. Since about 0.10 mL of DIP reagent must be added to produce a pink endpoint even if no vitamin C is present, subtract 0.10 mL from the average volume calculated in step 1 to determine the actual volume of DIP reagent required in the titration.
3. Multiply the actual DIP reagent volume from step 2 by the vitamin C–DIP reagent conversion factor (from the reagent bottle label). This calculation gives the average experimentally determined amount of vitamin C in the titrated fruit juice–MAA samples.
4. Now determine the experimental value for the amount of vitamin C in one 8-oz serving of the fruit juice. Since you only titrated 5.00 mL of fruit juice, and one serving contains 240 mL, multiply your step 3 result by the factor 240/5.
5. Finally, determine the experimental result for the percent of adult daily value (%DV) in one serving of the fruit juice. Use the nutritional information from the label, and the adult daily value (DV) of vitamin C.
6. Calculate the percent difference between the experimental determination and the claimed vitamin C content from the fruit juice label.

Refer to Data Analysis section of your Report Sheet on pp. 127–128.

NAME (print) ____________________________________ DATE ______________
LAST FIRST

LABORATORY SECTION ______ PARTNER(S) ______________________________

REPORT SHEET

PRE-LAB ASSIGNMENT

Pre-Lab Questions

1. List at least four foods or beverages (other than fruit juice) that contain vitamin C, and state their vitamin C content per serving.

2. Containers of vitamin C generally are dark brown, and their labels usually state that the bottle should be kept tightly capped. Explain why this is done.

3. Supplements containing vitamin C often tout its antioxidant properties and often claim that it helps to protect against cell damage and premature aging. Explain the rationale for this claim in terms of the chemistry involved.

NAME (print) ______________________________________ DATE ____________________
LAST FIRST

LABORATORY SECTION ______PARTNER(S) ______________________________________

Titration Data and Observations:

Brand/type of juice: ______________________________________

Description of juice: ______________________________________

Claimed percent DV for an 8-oz serving from the label: ____________________

DIP Reagent Volumes:	Trial 1	Trial 2	Trial 3
Final burette reading	______	______	______
Initial burette reading	______	______	______
DIP volume used	______	______	______

Describe what you observed during the titration of the fruit juice sample with DIP reagent.

DATA ANALYSIS

Vitamin C Titration Calculations and Analysis

Average volume used

____

Volume needed for color change with no vitamin C ____________

Actual DIP volume required for titration of vitamin C ____________

Conversion factor between vitamin C and DIP reagent ____________

Amount of vitamin C in titrated fruit juice sample

____

NAME (print) ______________________________________ DATE ________________
LAST FIRST

LABORATORY SECTION ______PARTNER(S) ______________________________________

Experimental amount of vitamin C in one serving ______________

Experimental percent of vitamin C DV from one serving ______________

Percent difference between experimental and claimed %DV ______________

SOURCES OF ERROR

Once the data have been collected, the possibility of error in the results must be addressed. Please cite at least three random sources of error and briefly explain how each source of error may have affected your data. For a review of error analysis, see *Error Analysis in the Chemistry Laboratory* in the Appendix at the end of this manual.

POST-LAB QUESTIONS

1. According to its label, V-8 juice contains 140% of the DV of vitamin C in each serving. Would the titration procedure you just used be a good choice to determine the vitamin C content of V-8 juice? Explain why or why not.

NAME (print) ______________________________________ DATE ____________________
LAST FIRST

LABORATORY SECTION ______ PARTNER(S) ________________________________

2. Compare your experimental determination of the vitamin C %DV with the %DV claimed by the fruit juice producer. Is your experimental determination close to the producer's claim, or is it much larger or smaller? Can you give a likely explanation for any large discrepancies?

CONCLUSION

Based on your analysis, will you recommend that the FDA validate the claimed vitamin C content in the fruit juice you tested? State your recommendation and the reasoning behind it in complete sentences. Be sure that your conclusion is complete and concise.

Experiment 14
Urinalysis

PRE-LAB ASSIGNMENT

Reading:

1. Read the experiment.
2. Refer to Chapters 10 and 16 of *Adventures in Chemistry*.

Question:

3. Sketch a full pH scale on your Report Sheet.
4. Place at least six commonly known liquids (e.g., milk, juice, etc.) in the appropriate place on your scale.
5. Provide complete reference(s) for source(s).

INTRODUCTION

A plastic soda bottle containing a yellow liquid was found by a security officer making her rounds in a closed, under-construction part of the local library. Not knowing what the yellow liquid was, the security officer requested help in the identification of this mysterious yellow liquid. Members of the Evidence Collecting Division (EDC) of UniChem International quickly identified the yellow liquid as urine. Since this portion of the library is closed to the public and library personnel, this left only members of the construction crew as possible suspects. When the foreman of the construction company was questioned, he provided the security officer with a list of the three workers who were assigned to that location. The foreman asked each of the three workers if they left or noticed the soda bottle containing the urine. All workers denied leaving the soda bottle and denied any knowledge of the soda bottle. When asked, all workers volunteered to give a urine sample. The urine samples were collected from each worker by an investigator from UniChem International and labeled as follows:

Evidence Sample 1 from Suspect 1
Evidence Sample 2 from Suspect 2
Evidence Sample 3 from Suspect 3
Evidence Sample 4 from Unknown Suspect

The evidence samples were stored properly to prevent sample degradation and taken back to the lab for analysis. Once the analysis is complete, this will determine if one of the three construction workers left the bottle of urine at the construction site.

We will work in teams of three investigators. Each team will be given a sample of urine from each suspect and a sample of the urine from the unknown suspect to analyze. Each team will need to use deionized water for a "blank" to ensure quality control. To perform a standard urinalysis will require us to measure **specific gravity** and **pH** and to test for the presence of **electrolytes**, **glucose** and **albumin** in each sample. The composition of urine will vary from individual to individual. Diet, fluid intake, and level of physical activity are some of the factors that influence the character of urine.

Urine is approximately 95% water; therefore, it will have a specific gravity close to 1. Specific gravity, also known as **relative density**, is a measure of the density of a material. For our purposes, specific gravity is equal to the density of the material divided by the density of water. It has no units. From our earlier work, you know that the density of water is 1 g per cubic centimeter (or 1 g/mL). The specific gravity has the same value as the density measurement, but without the unit. The specific gravity is measured with a **hydrometer**. A person who drinks more water than another person may have a urine sample with a specific gravity closer to 1.

The other ~5% of urine consists of ions that are eliminated by the body to help maintain proper levels of electrolytes and pH and other processes that take place. We will test for the common ions excreted by the body. These are sodium ion (Na^+), chloride ion (Cl^-), and sulfate ion (SO_4^{2-}).

Urine also contains urea and uric acid which both are produced by the body when substances in the body are broken down. Sometimes amino acids (trace amounts) and electrolytes can be found in urine. Any detectable acid can be measured with a pH meter or pH paper because acid produces hydrogen ions (H^+), in solution.

In a diabetic, the urine may contain glucose. The cells of diabetics cannot use all the glucose in the blood. The extra glucose accumulates in the bloodstream and is removed via the urinary system. The presence of glucose can be detected when urine turns from blue to yellow on addition of Benedict's solution. Reagent strips, such as Multistix or Clinitest, also can be used to detect the presence of glucose in urine. A range of colors from green to brown is indicative of a positive glucose test.

Glucose: "a sugar, $C_6H_{12}O_6$,...occurs widely in nature and is the usual form in which carbohydrate is assimilated by animals". —from *Webster's Dictionary*.

Albumin, water-soluble proteins, also may be present in the urine. On heating, the albumin will coagulate (becomes thickened, curdled, and/or clotted). Reagent strips also can be used to detect the presence of proteins in urine. Albustix or Multistix will produce yellowish-green to bluish-green color if proteins are present.

Albumin: "any of a number of heat-coagulable water-soluble proteins that occur in blood plasma or serum, muscle, the whites of eggs, milk, and other animal substances and many other plant tissues and fluid". —from *Webster's Dictionary*.

LEARNING OBJECTIVES

Be able to:

- Perform qualitative analysis tests on the collected urine samples.
- Use the information from the qualitative tests to identify the guilty party.

APPARATUS

Chemicals:

- Urine samples from Suspects 1, 2, and 3
- Evidence urine sample from library
- Dropper bottle of Benedict's solution
- Small bottle of 3 M HCl, hydrochloric acid (just for cleaning copper wire)
- Dropper bottle of 3 M hydrochloric acid (HCl)
- Dropper bottle of 3 M nitric acid (HNO_3)
- Dropper bottle of 0.1 M silver nitrate ($AgNO_3$)
- Dropper bottle of 0.1 M barium chloride ($BaCl_2$)
- Squirt bottle of deionized water

Equipment:

- Hydrometer
- Analytical balance
- pH meter or pH paper
- Permanent ink marker
- Box Kim wipes
- Round test tube ring rack for 250-mL beakers
- 13 x 100 mm test tubes
- Test tube rack
- Roll of colored tape (for labeling)
- 1000-µL automatic pipette
- Hot plate
- Bunsen burner with striker
- 6-in. piece copper wire with small loop on one end and attached to a cork at the other end.
- Package of Multistix Reagent Strips (optional)

- Are your safety glasses on?
- Hydrochloric and nitric acids are strong acids that may irritate the skin and clothing. Use care when handling these reagents. If you spill one of these acids, alert your supervisor immediately.
- Tie back long hair and roll up loose sleeves before lighting the Bunsen burners.
- Use care when handling beakers and test tubes used for and in hot-water baths.

Clean-up: 

- All untreated urine solutions can go down the drain with water.
- Collect all solutions from the chloride ion test and sulfate ion test in a container labeled as "Halogen/Metal Waste."
- Multistix Reagent Strips can be disposed of in the trash receptacle.
- Wash your hands thoroughly with soap and water before you leave the laboratory.

PROCEDURE 1

Physical Appearance Test:

1. Label five 150-mL beakers as follows: Suspect 1, Suspect 2, Suspect 3, Unknown Urine Sample and Blank.
2. Measure ~100 mL of each of the urine samples taken from the three suspects and the unknown urine sample into the pre labeled 150-mL beakers.
3. Measure ~100 mL of deionized water into the properly labeled beaker.
4. Describe the color of each urine sample and the blank.
5. Record your observations on your Report Sheet.

Refer to Procedure 1 section of your Report Sheet on p. 140.

PROCEDURE 2

Specific Gravity Test (with Hydrometers):

1. Using the urine sample from Beaker 1 (procedure 1), fill up (nearly to the top) a 100-mL hydrometer cylinder with the first urine sample to be tested (Suspect 1).
2. Slowly lower the hydrometer into the center of the cylinder.
3. Give the hydrometer a starting spin that will continue as a free-spinning motion.
4. Once the hydrometer stops spinning, read the incremented values on its upper glass stem.
   - ➢ To take an accurate reading, look for the value, that lines up with the uppermost water level.
   - ➢ The glass stem of the hydrometer has increments of specific gravity from 1.000 to 1.010 given in 10 equal increments.
   - ➢ You can record a specific gravity reading for any liquid between 1.000 and 1.060, to the nearest thousandths place with your hydrometer.
5. Record the specific gravity of the urine sample from Suspect 1 onto your Report Sheet.
6. Return the urine from the hydrometer to Beaker 1.
7. Rinse the hydrometer and the cylinder with distilled water.
8. Repeat the procedure to determine the specific gravity for urine samples taken from Suspect 2 and Suspect 3, the unknown urine sample, and the blank.
9. Record the specific gravity of each urine sample on your Report Sheet.
10. Properly dispose of all urine samples and rinse all beakers with distilled water for the next test.

Specific Gravity Test (without Hydrometers):

1. If hydrometers are not available, specific gravity may be calculated using the following procedure.
2. Measure 5 mL of the urine sample taken from Suspect 1 using a 10-mL graduated cylinder.
3. Pre-weigh a 25-mL beaker.
4. Record this value on your Report Sheet.
5. Transfer 5 mL of urine from the graduated cylinder to the pre-weighed beaker.
6. Using the same analytical balance, re-weigh the beaker and the urine.
7. Record this value on your Report Sheet.
8. Calculate the density of the urine taken from Suspect 1. Remember that density is defined as mass (g) per volume (mL).
   - To calculate the density, divide the mass of the liquid by the volume of the liquid. The unit for density is grams per milliliter.
9. Record this value on your Report Sheet.
   - Specific gravity is defined as the density of the substance (g/mL) per density of water (g/mL) (see equation below)

$$\text{Specific gravity} = \frac{\text{density of substance (g/mL)}}{\text{density of water (g/mL)}}$$

   - Specific gravity is recorded without units.
   - The density of water is defined as 1.00 g/mL.
   - Therefore the specific gravity of the urine sample from Suspect 1 is the same as the density without the units.
10. Record the value of specific gravity for the urine sample taken from Suspect 1 on your Report Sheet.
11. Return the urine to Beaker 1.
12. Repeat this procedure for the urine samples taken from Suspect 2, Suspect 3, the unknown urine sample, and the blank.
13. Use a clean, dry 25-mL beaker for each test.
14. Record the values on your Report Sheet.
15. Properly dispose of all urine samples and rinse all beakers with distilled water for the next test.

Refer to Procedure 2 section of your Report Sheet on p. 141.

PROCEDURE 3

pH Test (with pH Meters):

1. Start with fresh urine samples by measuring ~50 mL of each urine sample taken from the three suspects and the unknown urine sample into the pre-labeled 150-mL beakers.
2. Calibrate the pH meter as per instructions from your supervisor.
3. Lift the pH probe out of the storage solution, and rinse it off using a squirt bottle of deionized water.
4. The rinse water should fall into the beaker labeled “Rinse”.
5. Gently blot off any excess liquid at the bottom of the probe using a Kim wipe.

6. Lower the pH probe into the urine sample in Beaker 1 taken from Suspect 1.
7. Follow the instructions given by your supervisor as to how to use the pH meter.
8. Record pH value of the urine sample from Suspect 1 on your Report Sheet.
9. Remove the probe from the urine sample, and rinse off the probe as instructed in steps 2–3.
10. Set Beaker 1 aside.
11. Repeat this above procedure with the urine samples from Suspect 2, Suspect 3, the unknown urine sample, and the blank.
12. Record the pH of these samples on your Report Sheet.
13. Properly dispose of all urine samples and rinse all beakers with distilled water for the next test.

pH Test (without pH Meters):

1. Start with fresh urine samples by measuring ~50 mL of each urine sample taken from the three suspects and the unknown urine sample into the pre-labeled 150-mL beakers.
2. Obtain a roll or strips of pH paper and tear off a piece about 1 in. in length.
3. Dip a glass stirring rod into Beaker 1 containing the urine sample from Suspect 1, and then dab the piece of pH paper.
4. Compare the color of the tested pH paper with the color chart on the pH paper container.
5. Record the pH of the urine sample from Suspect 1 on your Report Sheet.
6. Rinse the glass stirring rod with distilled water, and dry it with a Kim wipe.
7. Repeat the procedure with the urine samples from Suspect 2, Suspect 3, the unknown urine sample, and the blank.
8. Record the pH of these samples on your Report Sheet.
9. Properly dispose of all urine samples and rinse all beakers with distilled water for the next test.

Refer to Procedure 3 section of your Report Sheet on p. 141.

PROCEDURE 4

Electrolyte Test:
Sodium Ions Test (Na^+):

1. Prepare a copper flame test wire by sticking one end of an 8-in piece of into a cork and making a small loop on the other end.
2. Clean the loop end of the flame test wire by sticking it into a small bottle of 3 M HCl. Carefully blot off any excess HCl with a Kim wipe.

   ⚠ If you spill any HCl, alert your supervisor immediately.
3. Using a striker, light a Bunsen burner. Adjust the air vent until the color of the flame is blue. Adjusting the gas vent will allow you to change the height of the flame. The desired height is ~6–7 cm.
4. Start with fresh urine samples by measuring ~50 mL of each urine sample taken from the three suspects and the unknown urine sample into the pre-labeled 150-mL beakers.
5. Dip the loop end of the flame test wire into Beaker 1 containing the urine sample from Suspect 1.
6. Place the loop with urine sample directly into the flame.

7. Are sodium ions present? A bright orange-yellow flame is indicative of a positive sodium ion test.
8. Record the color of the flame on your Report Sheet.
9. Repeat steps 5–7 with the urine samples from Suspect 2, Suspect 3, the unknown urine sample, and the blank.
10. Record observations on your Report Sheet for all samples tested.

Chloride Ions Test (Cl^-):

1. Label five test tubes as follows: 1, 2, 3, U, and B which corresponds to Suspects 1, 2, and 3, and the unknown urine sample and the blank, respectively.
2. Using a 10-mL graduated cylinder, measure 3 mL of urine from sample 1 and transfer it to the test tube labeled 1.
3. Add 5 drops of 3 M nitric acid (HNO_3) and 5 drops of 0.1 M silver nitrate ($AgNO_3$) solution to the test tube containing the urine sample from Suspect 1.
   ⚠ If you spill any HNO_3, alert your supervisor immediately.
4. Record your observations for the urine sample from Suspect 1 on your Report Sheet.
5. Are chloride ions present? The formation of silver chloride (AgCl) as a white precipitate is an indication of a positive chloride ion test.
6. Dispose of all solutions from this test in the container labeled as "Halogen/Metal Waste."
7. Repeat steps 2–6 with the urine samples from Suspect 2, Suspect 3, the unknown urine sample, and the blank.
8. Record observations on your Report Sheet for all samples tested.

Sulfate Ion Test (SO_4^{2-}):

1. Label five test tubes as follows: 1, 2, 3, U, and B which corresponds to Suspects 1, 2, and 3 and the unknown urine samples and the blank, respectively.
2. Using a 10-mL graduated cylinder, measure 3 mL of urine from Beaker 1 and transfer it to the test tube labeled 1.
3. Add 5 drops of 3 M hydrochloric acid (HCl), and 5 drops of 0.1 M barium chloride ($BaCl_2$), solution to the test tube containing the urine sample from Suspect 1.
   ⚠ If you spill any HCl, alert your supervisor immediately.
4. Record your observations for the urine sample from Suspect 1 on your Report Sheet.
5. Are sulfate ions present? The formation of barium sulfate ($BaSO_4$) as a white precipitate is an indication of a positive sulfate ion test.
6. Dispose of all solutions from this test in the container labeled as "Halogen/Metal Waste."
7. Repeat steps 2–6 with the urine samples from Suspect 2, Suspect 3, the unknown urine sample, and the blank.

Refer to Procedure 4 section of your Report Sheet on p. 142.

PROCEDURE 5

Albumin Test:

1. Fill a 250-mL beaker with about 200 mL of water.
2. Place a test tube ring-shaped rack into the beaker.

3. Place the beaker on a hot plate set to high to heat the water.
4. Label five test tubes with a permanent ink marker as follows: 1, 2, 3, and U and B, which correspond to Suspects 1, 2, 3 and the unknown urine sample and the blank, respectively. Do not use tape for labels, because the tape will fall off in the water bath.
5. Add urine from Beaker 1 to the test tube labeled 1 until half full.
6. Repeat step 6 with the urine samples from Suspect 2, Suspect 3, the unknown urine sample, and the blank.
7. Record observations on your Report Sheet of what the urine samples look like before heating.
8. Set test tubes into water bath rack, and wait for the water to boil.
9. Record observations on your Report Sheet after test tubes have boiled for 2 minutes.
   - The heat will cause any proteins present to coagulate.
   - Are any proteins present?
10. Carefully remove the hot test tubes from the heat using a test tube holder, and place the test tubes into a test tube rack.
11. If Multistix Reagent Strips are available, check each urine sample for the presence of protein by dipping the strip into urine sample labeled 1 from Suspect 1.
12. Follow the instructions on the container for the correct time to read the reagent strips.
13. Compare the color of the strip (at the correct time) with the color chart on the container.
14. Record the color and observations on your Report Sheet.
15. Repeat the reagent strip test with the with the urine samples from Suspect 2, Suspect 3, the unknown urine sample and the blank.
16. Record the observations on your Report Sheet.
17. Save all of the heated samples for the next test.

Refer to Procedure 5 section of your Report Sheet on p. 142.

PROCEDURE 6

Glucose Test:

1. Label a 25-mL beaker as "B's Solution".
2. Pour a small amount of Benedict's solution into the beaker labeled "B's Solution".
3. Using a new tip on the pipette slowly take up 1000 μL (that's 1 mL) of Benedict's solution on the automatic pipette.
4. Follow instructions from your supervisor on the correct way to use an automatic pipette.
5. Add 1 mL of Benedict's solution to each of the four urine samples and the blank you are testing for glucose (these are the samples you saved from Procedure 5).
6. Change the pipette tip if you touch the sides of the test tube while adding the Benedict's solution.
7. All solutions should be a bluish color (like the Benedict's solution).
8. Return test tubes to the round test tube ring rack in the hot-water bath.
9. Wait 3 minutes, and then observe the color of the urine samples.
10. Record observations for all four urine samples and the blank on your Report Sheet.
11. Is glucose present? If the color of the urine sample changes from blue to a yellow-orange or gold color, this is an indication that glucose is present in the urine sample.

12. If Multistix Reagent Strips are available, check each urine sample for the presence of glucose by dipping the strip into one of the samples.
13. Follow the instructions on the container for the correct time to read the reagent strips.
14. Compare the color of the strip (at the correct time) with the color chart on the container.
15. Record the color and observations for all four urine samples and the blank on your Report Sheet.
16. Repeat the reagent strip test with the remaining urine samples from Suspect 2, Suspect 3, the unknown urine sample, and the blank.
17. Record the observations on your Report Sheet.

Refer to Procedure 6 section of your Report Sheet on p. 143.

NAME (print) ______________________________ DATE ______________
LAST FIRST

LABORATORY SECTION ______ PARTNER(S) ______________________________

REPORT SHEET

PRE-LAB ASSIGNMENT

Pre-Lab Questions

1. Sketch a full pH scale on your Report Sheet and place at least six commonly known liquids (e.g., milk, juice, etc.) in the appropriate place on your scale.

2. Provide complete reference(s) for your source(s).

PROCEDURE 1

Physical Appearance: Summarize your results for physical appearance in the table below

Sample	Appearance
Blank	
Suspect 1 urine sample	
Suspect 2 urine sample	
Suspect 3 urine sample	
Unknown urine sample	

NAME (print) ______________________________________ DATE ____________________
LAST FIRST

LABORATORY SECTION ______ PARTNER(S) ______________________________

PROCEDURE 2

Specific Gravity: Summarize your results for specific gravity in the table below. If a hydrometer was used, fill in the first column only. If a hydrometer was not used fill in the entire table.

Sample	Specific Gravity	Mass of Beaker (g)	Mass of Beaker and Urine (g)	Volume of Urine in Beaker	Density (g/mL)
Blank					
Suspect 1 urine sample					
Suspect 2 urine sample					
Suspect 3 urine sample					
Unknown urine sample					

Sample calculation for specific gravity if a hydrometer was not used:

PROCEDURE 3

pH: Record the pH value in the table below

Sample	pH
Blank	
Suspect 1 urine sample	
Suspect 2 urine sample	
Suspect 3 urine sample	
Unknown urine sample	

NAME (print) ________________________________ DATE ____________________
LAST FIRST

LABORATORY SECTION ______ PARTNER(S) ________________________________

PROCEDURE 4

Electrolytes: Summarize your results for the presence of electrolytes in the table below. Indicate whether or not the ion was present in the solution by writing Y for yes or N for no and record all observations in the space provided.

Sample	Sodium Ions (Na^+) Yes or No	Chloride Ions (Cl^-) Yes or No	Sulfate Ions (SO_4^{2-}) Yes or No
Blank			
Suspect 1			
Suspect 2			
Suspect 3			
Unknown urine sample			

PROCEDURE 5

Albumin Test: Summarize you results for the presence of albumin in the table below. If Multistix Reagent Strips were not used, put NA in this column.

Sample	Albumin Observations	Albumin Reagent Strip Observations	Are Proteins Present? Yes or No
Blank			
Suspect 1			
Suspect 2			
Suspect 3			
Unknown urine sample			

NAME (print) ______________________________ DATE ______________
LAST FIRST

LABORATORY SECTION ______ PARTNER(S) ______________________________

PROCEDURE 6

Glucose Test: Summarize you results for the presence of glucose in the table below. If Multistix Reagent Strips were not used, put NA in this column.

Sample	Glucose Observations	Glucose Reagent Strip Observations	Is Glucose Present? Yes or No
Blank			
Suspect 1			
Suspect 2			
Suspect 3			
Unknown urine sample			

SOURCES OF ERROR

Once the data have been collected, the possibility of error in the results must be addressed. Please cite at least three random sources of error and briefly explain how each source of error may have affected your data. For a review of error analysis, see *Error Analysis in the Chemistry Laboratory* in the Appendix at the end of this manual.

NAME (print) ____ DATE ____________________
LAST FIRST

LABORATORY SECTION ______ PARTNER(S) ____

POST-LAB QUESTIONS

1. Give the balanced chemical equation for the reaction between barium chloride ($BaCl_2$) and sodium sulfate (Na_2SO_4). One of the products is soluble in water while the other product is not. Based on your observations, which product is insoluble in water, as evident by the presence of the white precipitate?

2. As a part of your yearly physical examination, your doctor ordered a urinalysis test. Based on the knowledge you gained from this experiment, why did your doctor order this test?

3. You are the owner of a small local trucking company. The underwriter for your insurance requires that all potential drivers have a pre employment urinalysis and undergo random urinalysis testing. Do you think that this is a good idea? Explain.

NAME (print) ____ DATE ______________________
LAST FIRST

LABORATORY SECTION ______ PARTNER(S) ____

CONCLUSION

Based on your analysis, which suspect, if any, was responsible for leaving the bottle of urine at the construction site of the library? How can you support your position with your data? What other things can you conclude or state? State your conclusions and the reasoning behind them in complete sentences. Be sure that your conclusion is complete and concise.

Experiment 15

Synthesis and Purification of Acetylsalicylic Acid

PRE-LAB ASSIGNMENT

Reading:

1. Read the experiment and appropriate sections from Chapters 6 and 11 of *Adventures in Chemistry*.

Research:

2. Look up the chemical structures for salicylic acid, acetic anhydride, and acetylsalicylic acid and record the structures on p. 159.
3. Circle the functional groups in each of the compounds listed in #2.
4. Record the molecular weights for each of the compounds listed in #2. The molecular weights may be looked up in a reference book or calculated.
5. Look up the melting points for salicylic acid and acetylsalicylic acid.
6. Provide a complete reference(s) for your source(s).

INTRODUCTION

The 110th annual high school science fair has ended in controversy. It appears that the winner of the science fair may have cheated. The winning project was the synthesis of acetylsalicylic acid, or aspirin, via an esterification reaction. The winner is accused by the second-place finisher of substituting in commercially available aspirin instead of synthesizing aspirin. The chairperson of the science fair has called on the expertise of the Organic Synthetic Department (OSD) at UniChem International to investigate this allegation.

An **esterification reaction** occurs when a carboxylic acid reacts with an alcohol in the presence of an acid catalyst according to the reaction below:

Leaves as water

Benzoic Acid (Carboxylic acid) + Methanol (Alcohol) $\xrightarrow{H_2SO_4}$ Methyl benzoate (Ester) + H_2O

Notice in the reaction that when the carboxyl group of the acid combines with the hydroxyl group of the alcohol, the products are the ester and water. Esterification reactions usually require a catalyst, such as sulfuric acid. The esterification reaction between a carboxylic acid and an alcohol is quite slow even in the presence of a catalyst. In order to speed up the rate of the reaction, an acid anhydride such as acetic anhydride is used instead of the carboxylic acid, which is shown in the reaction below. Acetic anhydride works better than benzoic acid and requires less of the catalyst. Acetylsalicylic acid can be synthesized by the following reaction:

$$\text{Salicylic acid (Alcohol)} + \text{Acetic Anhydride (Acid anhydride)} \xrightarrow{H_2SO_4} \text{Acetylsalicylic acid} + \text{Acetic Acid}$$

Salicylic acid was discovered in the eighteenth century by extracting it from its naturally occurring matrix, the bark of the willow tree. Salicylic acid was found to have antipyretic, analgesic, and anti-inflammatory properties and therefore was used to relieve fevers and reduce pain and inflammation. It also proved to have severe side effects because of its acidic nature pH ~5. This caused damaged to the lining of the mouth and esophagus and irritated the stomach. In addition to this, salicylic acid has a bitter taste. The functional group thought to be responsible for the side effects is the phenol group (hydroxyl group bonded to the aromatic ring). In the late 1800s, Felix Hofmann, a German chemist, thought that converting the hydroxyl group to the ester functional group or converting salicylic acid to acetylsalicylic acid, hopefully would decrease the harmful side effects of salicylic acid. Acetylsalicylic acid is less of an irritant than salicylic acid, but notice that it still contains the carboxylic acid functional group (—COOH). The presence of this functional group still may cause irritation to the lining of the stomach even when normal amounts are digested. The addition of buffers and binders tends to reduce the side effects.

The OSD at UniChem International have been asked to reproduce the experimental results of the winner of the science fair by following the winner's procedure given below. Once the experiment is complete, comparison of the aspirin synthesized by the OSD at UniChem International with the winner's aspirin and with commercially available aspirin will either vindicate or disqualify the winner.

The crude aspirin product will be recrystallized from ethanol. This is necessary to remove any unreacted salicylic acid. The purity of the aspirin products will be determined using melting points, the ferric chloride chemical test, and thin-layer chromatography. Finally, the percent yield of the reaction will be calculated to determine reaction efficacy.

Recrystallization:

Recrystallization is a commonly used technique by organic chemists to purify solid compounds by removing impurities. The major impurity in your synthesized aspirin is unreacted salicylic acid. Dissolving the crude aspirin sample in hot ethanol and water will cause salicylic acid to remain in solution and the pure aspirin to crystallize out of the solution. The pure aspirin sample will be collected by vacuum or gravity filtration.

Aspirin Purity Test:

Both the crude aspirin sample and the purified aspirin sample are not pure enough to digest for any headache this experiment may be causing you. Therefore, **DO NOT** consume any of the aspirin samples produced in this laboratory. The purity of your crude, purified, and commercially available aspirin samples will be tested using a 1% ferric chloride ($FeCl_3$) solution. Ferric chloride is a chemical that will test for the presence of the phenol functional group (—OH group bonded to an aromatic ring, such as benzene). Ferric chloride solution is a pale-yellow color. The presence of the phenol functional group will produce a color change. If the phenol group is present in the molecule, the color of the ferric chloride solution will change from yellow to purple, blue, green or red color. The intensity of the color is an indication of the amount of the phenolic compound present. If the color does not change or changes to brown or beige, this is indicative of a negative ferric chloride test.

Thin-Layer Chromatography (TLC):

TLC is a technique used in both industry and academic research as a way of separating the components of a mixture. TLC is able to separate the components of a mixture by exploiting the interactions of each component with the solvent and with the TLC plate and the molecular weight of each component in the mixture.

The TLC plate usually is a piece of glass or plastic coated with a thin layer of silica gel or alumina, which is used as the absorbent stationary phase. We will be using plastic plates coated with silica gel that contain a fluorescent indicator. A baseline is drawn on the plate using a pencil 1.0 cm from the bottom edge of the plate. A second line is drawn on the plate 4 cm above the baseline. This is called the **solvent front**. Small amounts of the compound to be tested that have been dissolved in an appropriate solvent are spotted on the baseline of the TLC plate using a small micropipette. The spotted plate then is placed in the developing chamber containing the mobile phase. The mobile phase is a chosen solvent that will move up the plate carrying the components of the substance with it. The more soluble the component is in the solvent, the further up the plate it will be carried. Components that are not very soluble in the solvent will remain near the bottom of the plate. Different components of the mixture can be determined based on how far they travel up the TLC plate.

Once the solvent or mobile phase has traveled to the solvent front, it is removed from the developing chamber and allowed to dry. The spots on the plate can be detected using an ultraviolet (UV) light or an iodine chamber. The components of the mixture will appear as dark-green spots under the UV light or as dark orange-brown spots with the iodine chamber.

The distance each spot moves can be calculated and this is known as the ***R_f* value**. R_f stands for **retention factor** and is determined by dividing the distance the component travels by the distance the solvent travels as shown by the equation below:

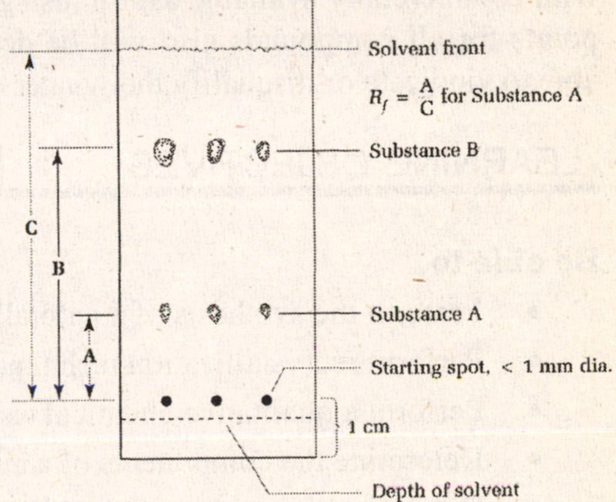

Figure 15.1 Developed TLC plate

$$R_f = \frac{\text{Distance the component moves}}{\text{Distance the solvent moves}}$$

The components of the mixture can be identified if its R_f values match the R_f values of the components of a known sample (see Figure 15.1).

Percent Yield Calculations:

The **percent yield** is an indication of the overall success of the reaction. The goal of any synthesis is to make as much product as possible. The percent yield of the reaction can be determined by dividing the actual amount of product synthesized by the theoretical yield of product times 100%. The **theoretical yield** is the maximum amount of product that can be produced if the reaction goes to completion. This rarely happens so the percent yield usually is less than 100%. The theoretical yield is determined by the amount of starting material present in the smallest molar quantity used in the synthesis. This starting material is called the **limiting reactant**. The limiting reactant in our synthesis of aspirin is salicylic acid.

Suppose that you started with 5.00 g of salicylic, the maximum amount of product that can be produced is determined by the following equation:

$$5.00\text{ g salicylic acid} \times \frac{1\text{ mol salicylic acid}}{138.12\text{ g salicylic acid}} \times \frac{1\text{ mol aspirin}}{1\text{ mol salicylic acid}} \times \frac{180.15\text{ g aspirin}}{1\text{ mol aspirin}}$$

$$= 6.52\text{ g aspirin} = \text{theoretical yield}$$

After completing the reaction you determine that you actually synthesized 5.62 g of aspirin. The percent yield of the reaction can be found using the following equation;

$$\text{Percent yield} = \frac{\text{Actual yield of product}}{\text{Theoretical yield of product}} \times 100 = \frac{5.62}{6.52} \times 100 = 86.2\%$$

In this experiment, you and your team of organic chemists will reproduce the winner's result by synthesizing aspirin. The aspirin you synthesize will be compared with the winner's aspirin and with commercially available aspirin using the ferric chloride chemical test and TLC. The melting points for all compounds also will be determined. Based on the data you collected you will be able to vindicate or disqualify the winner of the 110th annual high school science fair.

LEARNING OBJECTIVES

Be able to:

- Perform the synthesis of a naturally occurring compound.
- Perform recrystallization technique to purify product.
- Perform a qualitative chemical test and melting points to determine purity.
- Determine the components of a mixture using TLC.
- Determine the percent yield of a chemical reaction.

APPARATUS

Chemicals:

- Salicylic acid
- Acetic anhydride
- Concentrated sulfuric acid
- 95% ethanol (ethyl alcohol)
- 1% ferric chloride solution
- 75% ethyl acetates–Hexanes solution
- 0.15% salicylic acid solution
- Commercially available aspirin
- Iodine chamber
- Pure acetylsalicylic acid

Equipment:

- Hot plate
- Water aspirator hook up
- Filter paper that fits Büchner or Hirsch funnel
- 13 × 100 mm test tubes
- TLC plates
- UV lamp (short wave at 254 nm)
- Ruler with metric scale
- Pencil
- Filter paper, 9–11 cm diameter
- Mortar and pestle
- Aluminum foil or plastic wrap
- Pair of tweezers or metal tongs
- Melting-point capillary tubes
- Melting-point apparatus
- Weighing paper or weighing boat
- Roll of colored tape for labeling
- Permanent marker
- Analytical balance

- Are your safety glasses on?
- The aspirin synthesized during this experiment is not to be taken internally.
- Acetic anhydride and sulfuric acid can burn skin. Please wear gloves or wash your hands after handling these compounds.
- Measure acetic anhydride and sulfuric acid in a fume hood or a well-ventilated room.
- Acetic acid is a by-product of this reaction. The vapors may irritate the eyes and nose. Do not breathe in the vapors.
- Ethanol is highly flammable. No open flames while ethanol is in use.
- Iodine in the iodine chamber will stain your skin. Please wear gloves and use tweezers or metal tongs when adding or removing plates to the chamber.
- UV radiation is harmful to the eyes. Do not look directly into the lamp.

Clean-up: 

- All aqueous filtrates can go down the drain with water.
- Excess ethanol and ethanol filtrates should be collected in a labeled flammable waste container in the hood.
- Destroying excess acetic anhydride by the addition of water should take place in the fume hood.
- Ferric chloride solution should be collected in a labeled waste container in the fume hood.
- Excess salicylic acid should be collected in a labeled solid waste container in the fume hood.
- Dispose of your aspirin products as per your supervisor's instructions.
- Wash your hands thoroughly with soap and water before you leave the laboratory.

PROCEDURE 1

Synthesis of Acetylsalicylic Acid:

1. Add ~250 mL of distilled water to a 400-mL beaker and set the beaker on a hot plate. Set the hot plate to a medium to medium-high setting.
2. Weigh a clean, dry 125-mL Erlenmeyer flask to the nearest 0.01g, and record this value on your Report Sheet.
3. Using weighing paper or a weighing boat measure out 2.10 g of salicylic acid to the nearest 0.01g, and record this value on your Report Sheet.
4. Transfer the salicylic acid to the pre-weighed 125-mL Erlenmeyer flask, re-weigh the flask to the nearest 0.01 g, and record this value on your Report Sheet.

5. Determine the mass of salicylic acid in the flask by difference (subtract the mass of the flask from the mass of the flask and salicylic acid). Record this value on your Report Sheet.
6. In a fume hood, carefully measure out 10 mL of acetic anhydride in a graduated cylinder.
   ⚠ If you spill any acetic anhydride, alert your supervisor immediately.
7. Slowly add the acetic anhydride to the flask containing the salicylic acid.
8. Swirl the flask to mix the contents, being careful not to get salicylic acid stuck to the sides of the flask.
9. In a fume hood, slowly add 10 drops of concentrated sulfuric acid, H_2SO_4 to the flask.
   - **CAUTION:** Both acetic anhydride and sulfuric acid can burn the skin and ruin clothing.
     ⚠ If you spill any H_2SO_4, alert your supervisor immediately.
   - If you get either of these chemicals on your skin, wash the affected area with copious amounts of water.
10. Swirl to mix or mix using a glass stirring rod. **DO NOT USE A METAL STIRRING ROD!**
11. Place the flask in the hot-water bath. Heat the mixture with stirring until the entire solid has dissolved. Once all the solid has dissolved continue to heat for an additional 10 minutes.
12. Remove the flask from the hot-water bath and allow the mixture to cool to room temperature. **SAVE** the hot-water bath for the recrystallization of the crude aspirin product in Procedure 3.
13. While your solution is cooling, prepare an ice-water (using distilled water) bath in a 400-mL beaker.
14. In a fume hood or well-ventilated room, slowly add 20–30 drops of distilled water to the flask with swirling or stirring. Be careful not to breathe in the acetic acid vapors.
15. Slowly add an additional 50 mL of distilled water.
16. Place the flask in an ice-water bath for 10 minutes. You should start to see aspirin crystals fall out of the solution. (Save the ice-water bath for Procedure 3.)
17. If no crystals appear, it may be necessary to induce crystallization by gently scraping the sides of the flask with a stirring rod or metal spatula.

Refer to Procedure 1–3 Data Section of your Report Sheet on pp. 159-160.

PROCEDURE 2

Collection of Aspirin Crystals:

1. While you are waiting for the crystallization process to be completed, set up a filtration apparatus similar to the one shown in Figure 15.2.
2. Place a piece of filter paper (make sure you have a proper fit) into the funnel. Wet the filter paper with distilled water.
3. Turn on the water aspirator as far as it will go, and pour the aspirin crystals onto the filter paper. It may be necessary to push down on the funnel to make sure that you have a good seal.
4. Add ~10 mL of cold distilled water to the Erlenmeyer flask, swirl, and pour the water over the crystals in the funnel.

5. Move the crystals around with a spatula in order to spread them out over the entire filter paper. Be careful not to tear the filter paper.
6. Wash the crystals two more times with 10 mL each ice-cold distilled water (water can be taken from the ice-water bath from step 16 of Procedure 1). Allow air to pass over the crystals for at least 5–10 minutes with gentle stirring. This will dry the crystals. Disconnect the rubber tubing from the sidearm of the filter flask. Turn off the water aspirator.
7. While the crystals are drying, weigh a medium-sized watch glass and record this value on your Report Sheet.

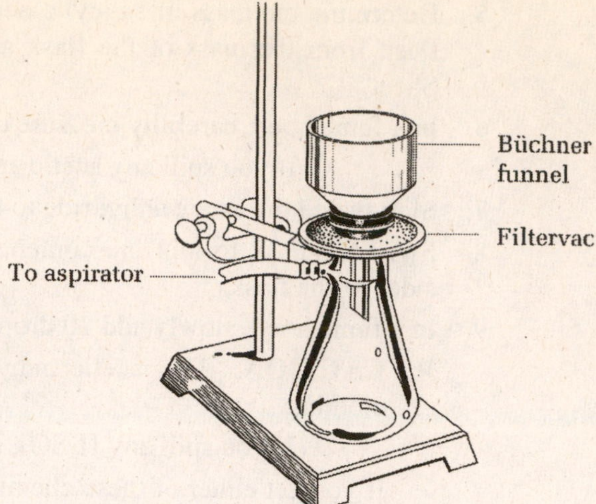

Figure 15.2 Vacuum filtration apparatus

8. Transfer the crystals to the watch glass. Make sure that the filter paper or parts of it do not transfer with the crystals. Re-weigh the watch glass and record this value on your Report Sheet.
9. Determine the mass of your crude aspirin product.
10. Calculate the percent yield of the reaction in the space provided on your Report Sheet.
11. Obtain a melting-point capillary tube and prepare a melting-point capillary of the crude aspirin sample by tapping the open end of the capillary tube into the crude aspirin sample on the watch glass.
12. Gently tap the closed end of the capillary tube against the benchtop to force the aspirin sample to the bottom of the tube. The capillary tubes may be dropped (bottom-end down) through a piece of glass tube that is resting on the benchtop or floor to aid in the packing process.
13. Set aside packed capillary tube containing the crude aspirin for later use.

Refer to Procedure 1–3 Data Section of your Report Sheet on pp. 159–160.

PROCEDURE 3

Recrystallization of Crude Aspirin:

1. Transfer the crude aspirin to a clean 125-mL Erlenmeyer flask.
2. Add 15 mL of 95% ethanol.
3. Place the flask into the hot-water bath (saved from step 15 of Procedure 1).
4. Swirl to dissolve the crystals.
5. Add ~30 mL of hot distilled water to the flask in the hot-water bath (can be taken from the hot-water bath).
6. Swirl to dissolve any remaining crystals. Keep the flask in the hot-water bath until all the crystals have dissolved.
7. Remove the flask from the hot-water bath and allow the solution to cool to room temperature.
8. While the mixture is cooling clean and reassemble the vacuum filtration apparatus.
9. Place the cooled mixture into an ice-water bath to maximize the recrystallization process.

As before, if no crystals form, scratch the side of the flask with a metal spatula.

10. Collect the recrystallized aspirin crystals with vacuum filtration. Remember to allow air to pass through the crystals to aid in the drying process.
11. Pre-weigh a watch glass and record this value on your Report Sheet.
12. Transfer the pure aspirin crystals to the watch glass and re-weigh.
13. Record this value on your Report Sheet.
14. Determine the mass of recrystallized aspirin product.
    - Do not be surprised if this amount is low.
18. Determine the percent recovery of recrystallized aspirin product using the equation below and record this value on your Report Sheet.

$$\text{Percent recovery} = \frac{\text{mass of recrystallized aspirin product}}{\text{mass of crude aspirin product}} \times 100$$

Refer to Procedure 1–3 Data Section of your Report Sheet on pp. 159–160.

PROCEDURE 4

Melting-Point Determination:

1. Obtain a sample of the science fair winner's recrystallized sample from your supervisor.
2. Prepare a melting-point capillary tube of your recrystallized aspirin product, pure acetylsalicylic acid, and the winner's recrystallized aspirin sample like you did in Procedure 2.
3. Get the capillary tube packed with your crude aspirin sample that you set aside in Procedure 2.
4. Follow the directions given to you by your supervisor for your particular melting-point apparatus.
5. Place the packed capillary tubes into the melting-point apparatus. Most melting-point apparatus can hold three to five capillaries at a time.
6. Turning on the instrument will light up the samples so that the solid material inside the capillary tubes is visible when looking through the magnifying eyeglass.
7. Set the dial on the apparatus to 5 or 6.
8. This will allow the apparatus to heat up quickly but will not cause the aspirin samples to melt.
9. Pure aspirin has a melting point of 138–140° C and salicylic acid melts at 158–161° C. The melting point of your crude aspirin sample is expected to be lower than that of the pure sample owing to the presence of impurities.
10. You will be able to see the temperature on the thermometer and the solid in the capillary tubes simply by raising your gaze up and down.
11. Allow the temperature to increase at a quick pace until the reading on the thermometer is close to 115–120 ° C.
12. The correct rate of heating while taking a melting point is 1–2° C per minute once you are close to the literature value of the sample. Adjust the rate of heating by turning the knob on the melting-point apparatus.
13. All melting points are recorded in a range. The range of the melting point starts at the first sign of melting and ends when the sample is melted completely. How close the

melting point of the sample under study is to the literature value of the sample and the size of the range are indications of how pure the sample is. A melting point that is far away from the literature value and a large space in the range of melting points are indications that the sample is not pure. A melting point that is close to the literature value with a small range is an indication of a pure compound.

14. Record the temperature on your Report Sheet for each sample at the first sign of melting and the temperature when the entire sample has melted.

Refer to Procedure 4 section of your Report Sheet on p. 160.

PROCEDURE 5

Ferric Chloride Test:

1. Obtain one tablet of commercially available aspirin. Record the type of aspirin on your Report Sheet. Indicate if the type of aspirin is buffered or not on your Report Sheet.
2. Using a mortar and pestle, crush the aspirin tablet until it is a fine powder.
3. Obtain six test tubes and numbered them as
   1. Pure salicylic acid
   2. 0.15% salicylic acid solution
   3. Crude aspirin sample
   4. Recrystallized aspirin sample
   5. Winner's recrystallized sample
   6. Commercially available aspirin
4. Place a few of the appropriate crystals into the labeled test tubes. Add 3 mL of distilled water to each of the test tubes.
5. Swirl gently to dissolve.
6. Add 2 drops of 1% ferric chloride solution to the test tubes one at a time.
7. Record the color of the solution and explain your observation on your Report Sheet. The formation of purple, blue, red, or green is an indication of a positive ferric chloride test, which indicates the presence of the phenol functional group.
8. Commercially available aspirin may contain up to 0.15% salicylic acid and still be considered pure. If the color of the aspirin sample is darker than the 0.15% standard salicylic acid solution, then it is not considered pure. If the color of the aspirin sample is lighter than the 0.15% standard salicylic acid solution, it is considered to be pure.
9. Compare the color of the 0.15% salicylic acid solution with the colors of your pure and crude aspirin samples.
10. Would your crude and recrystallized aspirin samples be considered pure or impure? How would you categorize the winner's aspirin sample?
11. Record your observations on your Report Sheet.

Refer to Procedure 5 section of your Report Sheet on p. 161.

PROCEDURE 6

Thin-Layer Chromatography (TLC):

1. Prepare the developing chamber by lining a 400-mL beaker with a piece of 9 × 11 cm piece of filter paper. The presence of the filter paper will ensure that the inside of the chamber is saturated with the eluting solvent. Add a little more than 1 cm of the eluting solvent, which is a 75% ethyl acetate–25% hexane solution to the beaker containing the filter paper.
2. **MAKE SURE THAT THE LEVEL OF THE ELUDING SOLVENT DOES NOT EXCEED 1 cm!**
3. Cover the beaker with aluminum foil or plastic wrap. Your developing chamber should look similar to the developing chamber pictured in Figure 15.3.

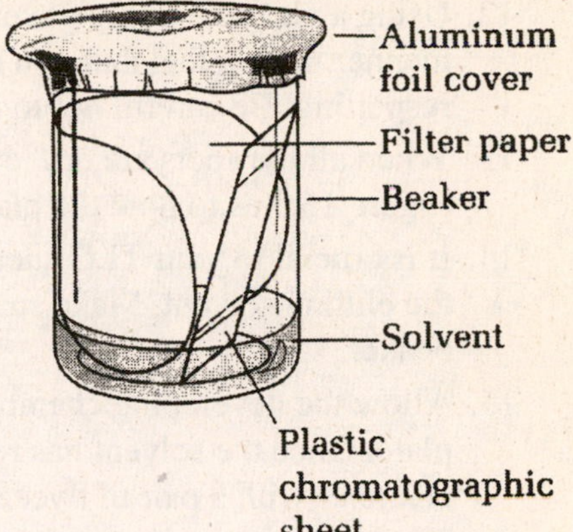

Figure 15.3 Development chamber

4. Obtain a precut plastic TLC plate from your supervisor. Be careful to only handle the plate from the sides. Notice that the plate has a shiny plastic side and a white-coated dual side. Do not touch the coated side of the plate because this may interfere with the components in your compounds.
5. Draw a line on the coated side of the plate 1 cm from the bottom with a pencil. This is the baseline.
6. Draw another line 4 cm above the baseline. This is the solvent front. Do not press too hard or the silica gel coating may come off.
7. Next, mark five small dots evenly spaced along the baseline. Labeled the dots 1 through 5 under the spot below the baseline. See Figure 15.4.
8. Obtain 5 test tubes and label them as follows:
    1. Pure Salicylic acid
    2. Crude aspirin
    3. Recrystallized aspirin
    4. Winner's aspirin
    5. Commercially available aspirin
9. Place a small amount of pure salicylic acid on the tip of a micro spatula and add it to the test tube labeled pure salicylic acid. Add 1 mL of 95% ethanol, and swirl to dissolve. All the solid may or may not dissolve. Repeat this procedure with the remaining four TLC samples.

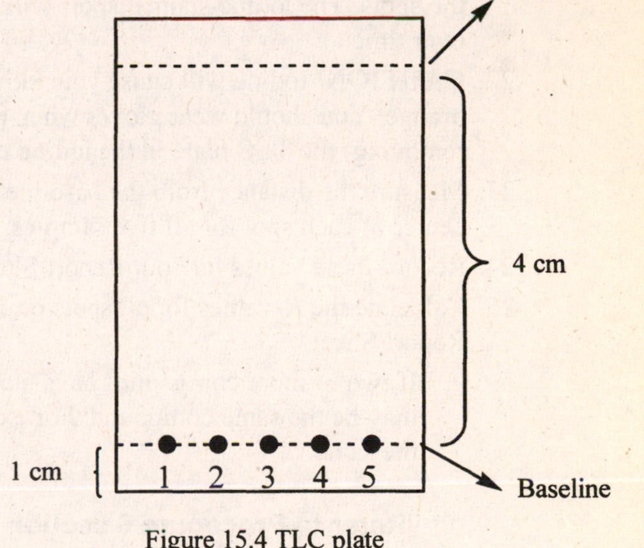

Figure 15.4 TLC plate before devolping

10. Dip a clean micropipette below the surface of the test tube 1. The liquid in the test tube will be drawn up into the micropipette by capillary action.

11. Lightly touch the tip of the micropipette to the first dot on the TLC plate. The dot should be as small as possible to minimize the overlapping of adjacent spots. You may find it necessary to do a few practice tries before actually attempting to spot the TLC plate.
12. Using a clean micropipette each time, spot the remaining TLC samples in the same manner. If you feel that you need to spot the same spot again, allow the spot to dry before respotting. Be careful not to mix up the TLC samples.
13. When all the spots are dry, carefully place the plate into the developing chamber (refer to Figure 15.3 as to how the plate should be placed inside the chamber).
14. It is crucial to your TLC success that the baseline of the TLC plate is above the level of the eluting solvent. Make sure to replace the aluminum foil or plastic wrap cover on the beaker.
15. Allow the developing chamber to stand undisturbed as the solvent moves up the TLC plate. Once the solvent has reached the solvent front, remove the TLC plate from the chamber with a pair of tweezers and allow the TLC plate to dry.
16. The developed TLC plate may appear the same after developing as it did before developing. Most spots on the TLC plate cannot be seen without the use of a UV lamp or iodine chamber. Lay the TLC plate flat on a clean surface with the coated side up.
17. Turn on the UV lamp and shine it directly on the TLC plate.
18. Circle all the spots you see with a pencil. See Figure 15.5.
19. If an iodine chamber is used to visualize the spots, place the TLC plate into the iodine chamber, and shake gently, if possible.
20. The components of the mixture will turn brown when exposed to iodine.
21. After several minutes, remove the TLC plate from the iodine chamber, and quickly circle all the spots. The iodine-stained spots will disappear over time.
22. **CAUTION:** Iodine will cause your skin to turn orange. You should wear gloves when placing (or removing) the TLC plate in the iodine chamber.
23. Measure the distance from the baseline to the center of each spot for all five samples.
24. Record these values on your Report Sheet.
25. Calculate the R_f values for all spots on the TLC plate. Record the R_f values on your Report Sheet.
    - ➢ If two or more compounds have similar R_f values, the probability is high that they may be the same compound. For example, in Figure 15.5, compounds 2 and 5 may be the same.

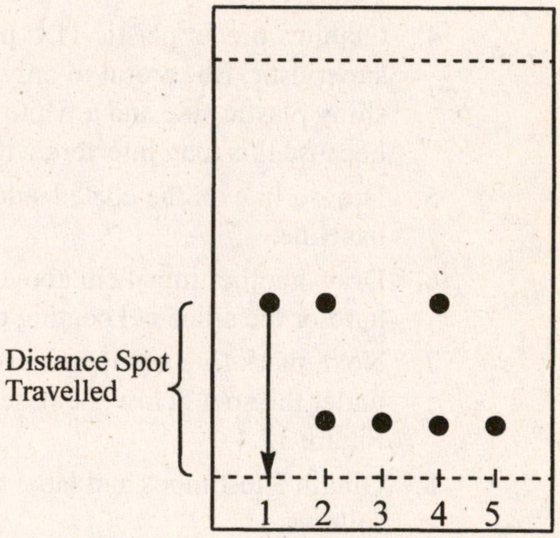

Distance Spot Travelled

1 2 3 4 5

Figure 15.5 TLC plate after devolping

Refer to Procedure 6 section of your Report Sheet on p. 161.

NAME (print) ______________________________________ DATE ____________________
LAST FIRST

LABORATORY SECTION ______ PARTNER(S) ______________________________________

REPORT SHEET

PRE-LAB ASSIGNMENT

Research

1. Look up the chemical structures for salicylic acid, acetic anhydride, and acetylsalicylic acid. Record the structures on your Report Sheet.

2. Circle the functional groups in each of the compounds listed in #1.

3. Record the molecular weights for each of the compounds listed in #1. The molecular weights may be looked up in a reference book or calculated.

4. Find the melting points for salicylic acid and acetylsalicylic acid.

5. Provide a complete reference(s) for your source(s).

Procedure 1–3 Data Section Synthesis and Recrystallization of Aspirin:

Mass of flask and salicylic acid __________g

Mass of flask __________g

Mass of salicylic acid __________g

Theoretical yield of aspirin __________g
Show calculations

Mass of watch glass and aspirin __________g

Mass of watch glass __________g

Mass of aspirin __________g

Percent yield __________%
Show calculations

NAME (print) ____________________________________ DATE ____________________
LAST FIRST

LABORATORY SECTION ______ PARTNER(S) ____________________________________

Mass of watch glass and recrystallized aspirin __________g

Mass of watch glass __________g

Mass of recrystallized aspirin __________g

Percent recovery __________%

Show calculations

PROCEDURE 4

Melting-Point Determination:

Sample	Start of Melting-Point Range	End of Melting-Point Range
Crude aspirin		
Recrystallized aspirin		
Winner's aspirin		
Pure salicylic acid		
Commercial aspirin buffered? Yes or no		

NAME (print) ____ DATE ____________________
LAST FIRST

LABORATORY SECTION ________ PARTNER(S) ______________________________________

PROCEDURE 5

Ferric Chloride Test:

Sample	Color of Ferric Chloride Solution	Explanation
Crude aspirin		
Recrystallized aspirin		
Winner's aspirin		
Pure salicylic acid		
0.15% Salicylic acid solution		
Commercial aspirin buffered? Yes or no		

PROCEDURE 6

Thin-Layer Chromatography (TLC):

Spot Number	Sample	Distance Moved	R_f Value
1	Pure salicylic acid		
2	Crude aspirin		
3	Recrystallized aspirin		
4	Winner's aspirin		
5	Commercial aspirin buffered? Yes or no		

NAME (print) ______________________________ DATE ______________
LAST FIRST

LABORATORY SECTION ______ PARTNER(S) ______________________________

SOURCES OF ERROR

Once the data have been collected, the possibility of error in the results must be addressed. Please cite at least two random sources of error and briefly explain how each source of error may have affected your data. For a review of error analysis, see *Error Analysis in the Chemistry Laboratory* in the Appendix at the end of this manual.

POST-LAB QUESTIONS

1. Sulfuric acid was used as a catalyst in the synthesis of aspirin. What is a catalyst?

2. How did the melting point of your recrystallized aspirin sample compare to the literature melting-point value for acetylsalicylic acid?

3. Was the melting point for your recrystallized aspirin closer to the literature value than your crude aspirin sample? Briefly explain.

NAME (print) ____ DATE ____________________
LAST FIRST

LABORATORY SECTION ______ PARTNER(S) ____

4. Aspirin that has been kept past its expiration date may start to smell like vinegar and gives a positive ferric chloride test. What chemical reaction might explain these observations?

CONCLUSION

Based on your data, what will you recommend to the chairperson of the 110th annual high school science fair? Using good chemical intuition, should the winner be vindicated or disqualified? What data will you refer to in order to support your findings? What other things can you conclude or state? State your conclusions and the reasoning behind them in complete sentences. Be sure that your conclusion is complete and concise.

Experiment 16

The Chemistry of Acid Rain

PRE-LAB ASSIGNMENT

Reading:

1. Read the experiment.

Research:

2. Research how acid rain forms, and the environmental effects of acid rain (some information is found in Chapter 15 of *Adventures in Chemistry*).

Question:

3. Answer the Pre-Lab Questions.

INTRODUCTION

One of the most serious environmental problems is that of acid deposition, primarily in the form of acid rain. All rain, even in completely unpolluted areas, is somewhat acidic owing to the carbonic acid (H_2CO_3) formed when atmospheric carbon dioxide dissolves in water. However, human activity has seriously increased the acidity of precipitation in many regions of the world.

Sulfur oxides, collectively known as $\mathbf{SO_X}$, are formed from the combustion of sulfur, which is found as an impurity in coal and gasoline. These SO_X gases then dissolve in water droplets to form sulfuric (H_2SO_4) and sulfurous (H_2SO_3) acids. Nitrogen oxides, known as $\mathbf{NO_X}$, are formed when atmospheric nitrogen and oxygen react in high-temperature conditions such as in a car engine or near a lightning strike. When NO_X gases dissolve in water droplets, they form nitric (HNO_3) and nitrous (HNO_2) acids.

The Environmental Protection Agency (EPA) is responsible for monitoring the environment in the United States, including such things as air and water pollution. Since the Environmental Chemistry Division (ECD) of UniChem International is recognized as one of the worldwide leaders in atmospheric chemistry research and remediation, the EPA has turned to the experts at UniChem to perform some research into acid rain formation and its effects.

You and your fellow members of the ECD team at UniChem International have been assigned several acid precipitation-related research projects. You will produce the gases CO_2, NO_2, and SO_2, and use small environmental chambers to simulate and analyze the dispersal of these gases in the atmosphere. You will investigate the effects of the acidic precipitation that is formed, as well as measures that might be taken to mitigate the effects of the acid rain. In addition, you will dissolve these gases in water to produce the various types of acidic rain and determine the effect of each gas on the pH of the rain.

LEARNING OBJECTIVES

Be able to:

- Describe the formation of acid rain using the appropriate reaction equations.
- Create simulated acid rain by producing gases and dissolving them in water.
- Measure the acidity of acid rain, and correlate precursor gases with pH.
- Study the formation of acid precipitation and ways to neutralize it.

APPARATUS

Chemicals:

- Universal indicator (UI), red cabbage juice, or other acid–base indicator
- Indicator/pH color scale
- 1.0 M sodium bicarbonate solution ($NaHCO_3$) in small plastic pipettes
- 1.0 M sodium nitrite solution ($NaNO_2$) in small plastic pipettes
- 1.0 M sodium bisulfite solution ($NaHSO_3$) in small plastic pipettes
- 1.0 M ammonia solution (NH_3) in small plastic pipettes
- 1.0 M hydrochloric acid (HCl) in small plastic pipettes
- Solid sodium bicarbonate ($NaHCO_3$)
- Solid sodium nitrite ($NaNO_2$)
- Solid sodium bisulfite ($NaHSO_3$)
- Boiled distilled or de-ionized water (pH 7)
- 6 M HCl solution in small plastic pipettes
- 6 M NH_3 solution in small plastic pipettes
- Lake water samples
- Plant samples (flower petals)

Equipment:

- Environmental reaction chamber (Petri dish and cover)
- Narrow-stem plastic pipettes
- pH meter

- Are you wearing your safety glasses?
- Perform all reactions in a well-ventilated room.

Clean-up: 

- Dispose of all excess reagents and waste as directed by your supervisor.
- Wash your hands thoroughly with soap and water before you leave the laboratory.

PROCEDURE 1

Producing and Dispersing NO_2, SO_2, and NH_3 Gases:

1. Use a small plastic pipette to place small drops of your chosen indicator on the inside of the cover of the Petri dish. Refer to Figure 16.1a. (These drops will simulate raindrops, and also register pH changes.)
2. Place a tiny amount of $NaNO_2$ solid (about the size of a pinhead) in the center of the bottom of a Petri dish. Refer to Figure 16.1b.
3. Place a drop of HCl solution on the solid $NaNO_2$. Quickly invert the Petri dish cover without dislodging the small drops of indicator, and place it over the Petri dish bottom. Refer to Figure 16.1c. (This creates the actual environmental reaction chamber.)

   ⚠ If you spill any HCl, alert your supervisor immediately.

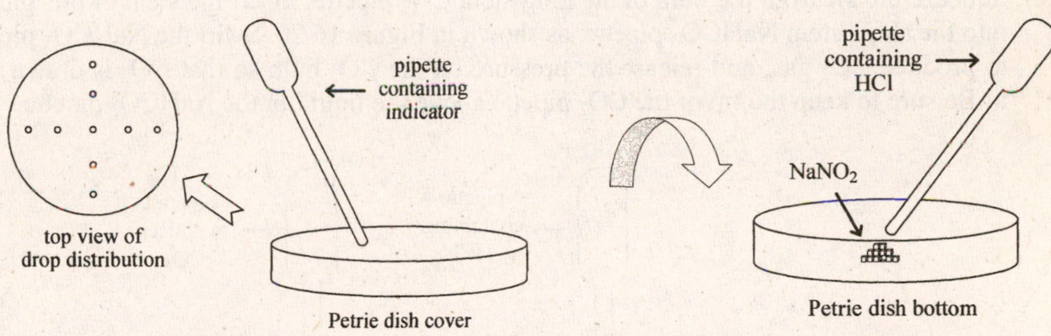

Figure 1.a. Figure 16-1.a.
Placing indicator drops

Figure 1.b.
Producing NO_2 gas

4. The reaction of HCl with $NaNO_2$ produces NO_2 gas which then diffuses throughout the environmental reaction chamber. Observe the drops of indicator, and note when they change color. Record the initial and final colors and pH values and the time needed for complete color/pH change for each drop.
5. Clean and dry the environmental reaction chamber.
6. Repeat steps 1–5, but use a small amount of solid $NaSO_3$ instead of $NaNO_2$.
7. Again repeat steps 1–5, but use one drop of NH_3 solution instead of the solid $NaNO_2$, and omit step 3. In other words, after placing the simulated raindrops on the cover, place the NH_3 solution in the bottom (but do not add HCl), place the cover on, and observe.

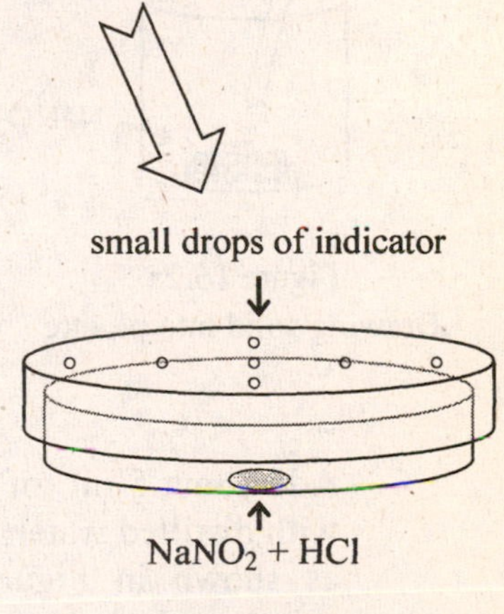

Figure 16.1c
Environmental
Reaction Chamber

Refer to Procedure 1 section of your Report Sheet on p. 171.

PROCEDURE 2

Measuring the pH of "Acid Rain":

1. Set up the pH system for data collection, as directed by your environmental chemistry research team director.
2. Construct three short-stem plastic pipettes, as directed. Label the short-stem pipettes with the formulas of the solids they will contain: $NaHCO_3$, $NaNO_2$, and $NaHSO_3$.
3. Make three narrow-stem pipettes, as directed. Label the narrow-stem pipettes with the formulas of the gases they will contain: CO_2, NO_2, and SO_2.
4. Squeeze the bulb of the $NaHCO_3$ pipette to force out the air. Place the end of the pipette into a small container of solid $NaHCO_3$, and release the bulb, drawing $NaHCO_3$ into the pipette until it fills the curved end of the bulb, as shown in Figure 16.2a.
5. Carefully insert the narrow stem of a plastic pipette containing 1.0 M HCl into the larger stem of the pipette containing solid $NaHCO_3$, as shown in Figure 16.2b. Gently squeeze the HCl pipette to add 20 drops of HCl solution to the solid $NaHCO_3$, and then remove the HCl pipette. Do not mix the $NaHCO_3$ and HCl until step 6.

   ⚠ If you spill any HCl, alert your supervisor immediately.
6. Squeeze the air from the bulb of the long-stem CO_2 pipette. Insert the stem of this pipette into the short-stem $NaHCO_3$ pipette, as shown in Figure 16.2c. Swirl the $NaHCO_3$ pipette to produce CO_2 gas, and release the pressure on the CO_2 bulb so that CO_2 is drawn into it. Be sure to keep the tip of the CO_2 pipette above the liquid in the $NaHCO_3$ pipette.

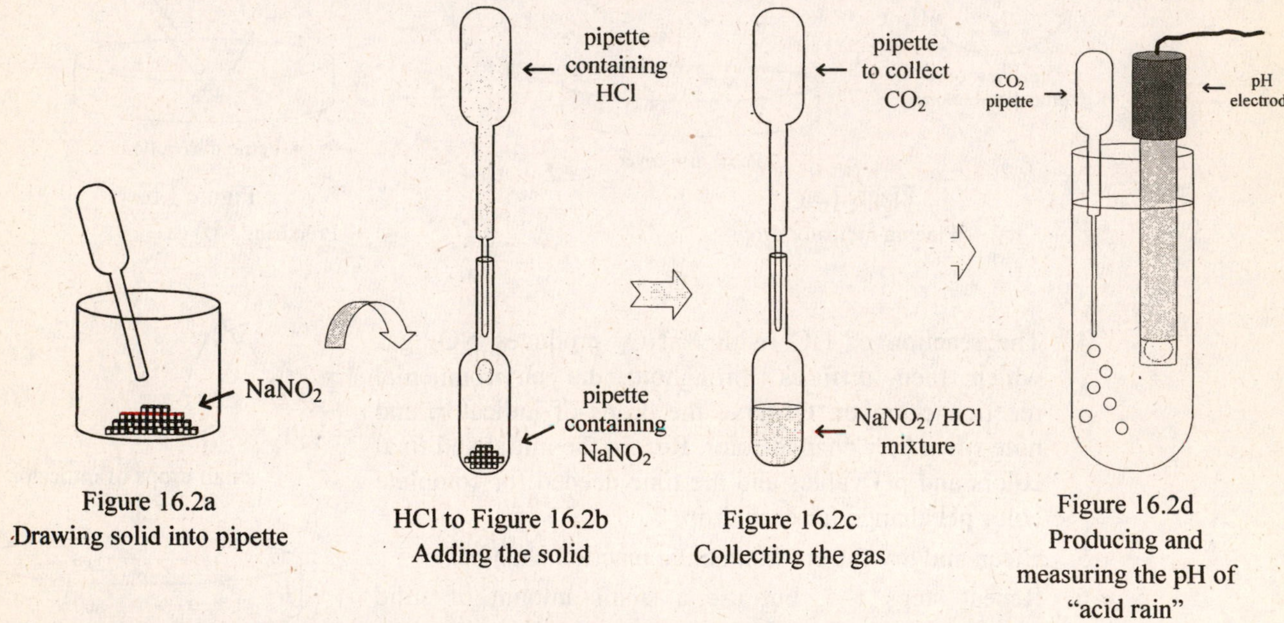

Figure 16.2a Drawing solid into pipette

HCl to Figure 16.2b Adding the solid

Figure 16.2c Collecting the gas

Figure 16.2d Producing and measuring the pH of "acid rain"

7. Add about 5 mL of boiled distilled water to a medium test tube. Rinse the pH electrode with distilled water, and carefully pat the tip dry. Place the pH electrode into the test tube as shown in Figure 16.2d. Make sure that the tip of the electrode is completely submerged in the water. When the reading has stabilized, record the initial pH of the water.
8. Insert the stem of the CO_2 pipette into the test tube next to the pH electrode so that its tip reaches the bottom of the test tube, as shown in Figure 16.2d.

9. Gently squeeze the pipette bulb to slowly bubble CO_2 through the water. Continue until all the CO_2 has been pushed out of the pipette bulb.
10. Monitor the pH for the next several minutes. Record the lowest pH reached after the CO_2 was added and the time it takes to reach the minimum pH. Be sure that you have monitored the pH for long enough to have definitely reached the lowest pH!
11. Repeat steps 4–10, but using solid $NaHSO_3$ instead of $NaHCO_3$ and the $NaHSO_3$ and SO_2 pipettes.
12. Repeat steps 4–10, but using solid $NaNO_2$ instead of $NaHCO_3$ and the $NaNO_2$ and CO_2 pipettes.
13. Thoroughly rinse the pH electrode between measurements, and return it to its storage solution when done, as directed by your research team director.

Refer to Procedure 2 section of your Report Sheet on p. 172.

PROCEDURE 3

Effect of Acid Precipitation on Simulated Lakes:

1. Place a small amount of solid $NaNO_2$ in the center of the bottom of the environmental reaction chamber.
2. Place one drop of each lake sample equidistant from the center of the environmental reaction chamber. Record the position of each lake sample.
3. Add one drop of indicator to each lake sample drop.
4. Add one drop of HCl solution to the solid $NaNO_2$ to produce NO_2 gas.
5. Quickly place the cover on the environmental reaction chamber. Observe the lake samples. Record the color changes that you observe and how long each takes to occur.

Refer to Procedure 3 section of your Report Sheet on p. 172.

PROCEDURE 4

Effect of Acid Precipitation on Plants:

1. Place a small amount of $NaHSO_3$ in the center of the bottom of the environmental reaction chamber bottom.
2. Place different plant samples equidistant from the center of the environmental reaction chamber. Be sure to include one undamaged flower petal and one that has been damaged by poking small holes in its surface. Record the position of each plant sample.
3. Add one drop of HCl solution to the solid $NaHSO_3$ to produce SO_2 gas.
4. Quickly place the cover on the environmental reaction chamber. Observe the plant samples. Record any changes that you observe and how long each takes to occur.

Refer to Procedure 4 section of your Report Sheet on p. 173.

NAME (print) ____________________ DATE ____________
LAST FIRST

LABORATORY SECTION ______ PARTNER(S) ____________________

REPORT SHEET

PRE-LAB ASSIGNMENT

Pre-Lab Questions

1. **Which human activities produce the largest quantities of each of the following gases?**

   a. CO_2

   b. NO_2

   c. SO_2

2. **Write the balanced chemical equations for the two sequential chemical reactions that describe the primary way that NO_2 gas is produced in the atmosphere.**

3. **Write the balanced chemical equations for the reactions described below that form the three types of acid precipitation that will be investigated in this experiment.**

   a. **Carbon dioxide gas reacts with water to form carbonic acid.**

   b. **Nitrogen dioxide gas reacts with water to form nitrous and nitric acids.**

   c. **Sulfur dioxide gas reacts with water to form sulfurous acid.**

NAME (print) ______________________________ DATE ______________
LAST FIRST

LABORATORY SECTION ______ PARTNER(S) ______________________________

PROCEDURE 1

1. Record the pH and time observations for the production and dispersal of CO_2 gas in the environmental reaction chamber.

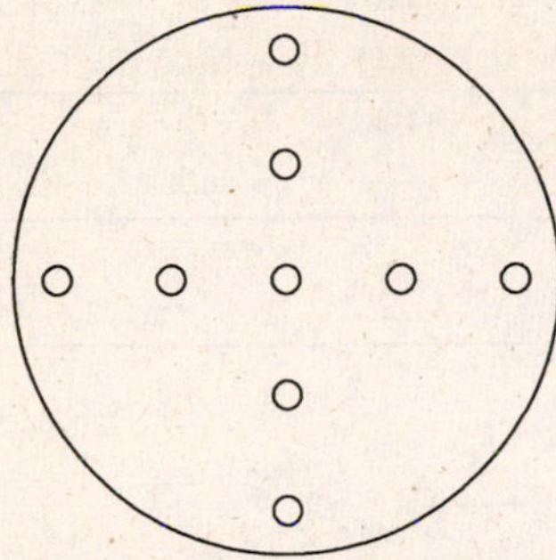

2. Record the pH and time observations for the production and dispersal of NO_2 gas in the environmental reaction chamber.

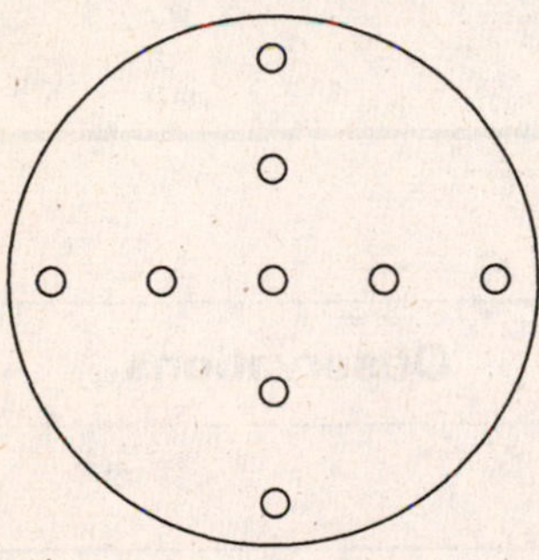

3. Record the pH and time observations for the production and dispersal of SO_2 gas in the environmental reaction chamber.

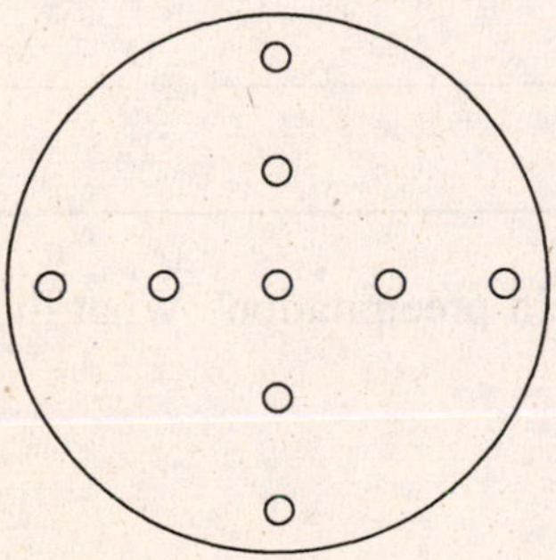

NAME (print) ______________________________ DATE ______________
LAST FIRST

LABORATORY SECTION ______ PARTNER(S) ______________________________

PROCEDURE 2

1. Record your observations.

Gas	Initial pH	Final pH	Time	Observations
CO_2				
NO_2				
SO_2				

2. Which gas had the smallest effect on pH?

3. Which gas had the largest effect on pH?

4. Which gas changed the pH most quickly?

PROCEDURE 3

1. Record your observations.

Type of Lake	Initial pH of Lake	Final pH of Lake	Time for Change	Observations

2. Which simulated lakes were most resistant to the effects of acid precipitation? What did these lakes have in common?

NAME (print) ______________________________ DATE ______________
LAST FIRST

LABORATORY SECTION ______ PARTNER(S) ______________________________

PROCEDURE 4

1. Record your observations.

Type of Plant	Observations

2. Which flower petal sample was most affected by the acid-producing gas?

SOURCES OF ERROR

Once the data have been collected, the possibility of error in the results must be addressed. Please cite at least two random sources of error and briefly explain how each source of error may have affected your data. For a review of error analysis, see *Error Analysis in the Chemistry Laboratory* in the Appendix at the end of this manual.

NAME (print) ________________________________ DATE ____________
LAST FIRST

LABORATORY SECTION ______ PARTNER(S) ____________________

POST-LAB QUESTIONS

1. Write the balanced equations for the reactions that were used in this experiment to produce the acid precipitation-forming gases CO_2, SO_2 and NO_2.

   a. When $NaHCO_3$ reacts with HCl, it forms carbon dioxide, water and sodium chloride.

   b. When $NaNO_2$ reacts with HCl, it forms sodium nitrate, sodium chloride, nitrogen monoxide, and water. Then the nitrogen monoxide reacts with oxygen gas to form nitrogen dioxide.

   c. When $NaHSO_3$ reacts with HCl, it forms sulfur dioxide, water and sodium chloride.

2. Refer to the Procedure 1 results to describe the general dispersal pattern of the acid precipitation-causing gases.

3. Refer to the Procedure 2 results to describe the effects of the gases CO_2, NO_2 and SO_2 on the acidity of acid rain.

4. Refer to the Procedure 3 results to describe the effects of acid deposition on the simulated lakes.

NAME (print) ____ DATE ______________________
LAST FIRST

LABORATORY SECTION ______PARTNER(S) ____

5. Refer to the Procedure 4 results to describe the effects of acid deposition on plants.

6. Which gas used in this experiment would cause rainfall even in unpolluted air to have a pH of less than 7?

   b. What acid is formed by this gas?

7. Which gas was produced in this experiment to

   a. determine the effects on the simulated lakes?

   b. determine the effects on plants?

8. Rainfall in the United States seldom falls below a pH of 4. Why would this simulation experiment produce lower pH values than those found in the environment? Explain.

9. Although human activities produce much more CO_2 than either NO_X or SO_X, there is much more concern about the contributions of NO_X and SO_X to acid rain. Explain why this is the case.

NAME (print) ____ DATE ______________________
LAST FIRST

LABORATORY SECTION ______ PARTNER(S) ____

CONCLUSION

Based on your research and experiments, summarize your findings on the dispersal of acid rain producing gases throughout the atmosphere, the effects of each type of gas on the acidity of the precipitation that is formed, and the effects of acid precipitation on lakes and plants. Answer these questions in complete sentences. Be sure that your conclusion is complete and concise.

Experiment 17

White Powder Identification

PRE-LAB ASSIGNMENT

Reading:

1. Read the experiment, and refer to Chapter 16 of *Adventures in Chemistry*.

Research:

1. Find the chemical formula for each of the following substances: baking soda, cornstarch, sugar, salt, benzoic acid, boric acid, calcium chloride, and plaster of paris.
2. Record the chemical formulas on your Report Sheet.
3. Provide complete references for where you found the formulas.

INTRODUCTION

A strange white powder has been discovered by the custodial staff in the girls' bathroom and in the boys' locker room at Dareigoto High School. During classroom change, Officer Look noticed a baggie containing a white powder sticking out of a student's coat pocket. On questioning, the student removed the baggie from his pocket, and the white powder spilled on the hallway floor.

Your unit of the Illegal Substance Division (ISD), at UniChem International has been called in to collect the dry white powder from the spill in the hallway, the bathroom and the locker room and to search the school for possible sources of the unidentified white powder. During the search, members of your unit collected eight white powders commonly used in the school. Baking soda, cornstarch, sugar, and salt are used in cooking classes by the home economics teacher. The chemistry teacher uses two white powders for experiments: benzoic acid and boric acid. The art teacher uses two white powders: calcium chloride (used for drying flowers) and plaster of paris (for sculpting).

The samples of all eight known powders, as well as the three samples of the white mystery powders from the bathroom, locker room, and hallway were properly labeled and have been transported back to the laboratories at UniChem International. The samples were labeled as follows:

Evidence Sample 1 White powder/girls bathroom
Evidence Sample 2 White powder/boys locker room
Evidence Sample 3 White powder/hallway
Sample 4 Baking powder
Sample 5 Cornstarch
Sample 6 Sugar
Sample 7 Salt
Sample 8 Benzoic acid
Sample 9 Boric acid
Sample 10 Calcium chloride
Sample 11 Plaster of Paris

The lab supervisor has already completed a chain-of-custody form to keep the evidence samples in our work area for three hours today. Each evidence sample has a unique identification label.

> A **Chain-of-Custody** document enables all forensic analysts working on a sample to keep track of where a sample is located and know who has worked with the sample and when it was worked on.

Working in pairs, you will develop the **qualitative** method needed to analyze and determine the identities of the three unknown white powders. Your supervisor will assign each pair of agents one of the white powders to analyze. The results of your analyses will be compiled with the results collected by the other members of your division to develop the qualitative method, which will be used to identify the unknown evidence samples.

> **Qualitative analysis** in the strict chemical sense has to do with designing a chemical test to allow the identification of the components of any given substance or mixture. In a broad context, this type of analysis typically involves a comparison of characteristics (e.g., color, shape, and consistency) that are not always quantifiable. In other words, the comparisons are not usually measured, merely observed.

LEARNING OBJECTIVES

Be able to:

- Learn the techniques of purification of solids, such as the white powder spilled during the search.
- Use qualitative analysis to identify the unknown white powders.

APPARATUS

Chemicals:

Small vials containing the following compounds:

- Baking soda
- Cornstarch
- Sugar
- Salt
- Calcium chloride
- Plaster of paris
- Benzoic acid
- Boric acid

Small dropper bottles containing the following:

- Acetic acid (vinegar)
- Tincture of iodine solution
- 70% isopropyl alcohol
- Small bottle of 1 M sodium carbonate solution
- Squirt bottles of deionized water

Equipment:

- Hand lens (used to examine the appearance of the white powders)
- 13 x 100 mm test tubes
- Test tube rack
- 10 corks that fit test tubes
- 10–15 disposable plastic pipettes
- Hot plate
- Sheet of colored paper (we need this for looking at the powders), sheet of unlined white paper for under the spot plate
- Piece of fluted filter paper
- Several pieces of filter paper to aid in the drying process
- Ring stand
- Utility clamp
- Analytical balances (shared equipment)
- Spot plate (see Figure 2.1)
- Roll of colored tape used for test tube labels
- Vortex mixer
- 8 aluminum foil heating dishes

- Are your safety glasses on?
- Do not taste substances in the lab, even if you think you know what they are.
- Work in the fume hood when dispensing acetic acid (vinegar).
- Sodium carbonate is basic and can irritate the skin. Wash your hands after using it or wear gloves.
- Isopropyl alcohol is very flammable; therefore, no open flames when using this substance.

Clean-up: 

- No special clean-up is necessary for this experiment. All liquid can go down the drain.
- Wash your hands thoroughly with soap and water before you leave the laboratory.

PROCEDURE 1

Purify Evidence Sample 3, the Spilled White Powder:

1. Add 150 mL of deionized water to a 250-mL beaker, and place it on a hot plate for a hot-water bath.
2. Heat the water in the beaker until hot. Adjust the setting on the hot plate to keep the water hot.

3. Weigh ~0.500 g of Evidence Sample 3 on a piece of tarred weighing paper. Record the weight of Evidence Sample 3 on your Report Sheet.
4. Transfer the weighed powder to a 13 × 100 mm test tube.
5. Using a plastic pipette, slowly add drops hot water (from the hot-water bath prepared in step 1) until all the white powder in the test tube dissolves.
6. Using a test tube clamp place the test tube in the hot-water bath to keep the water in the test tube hot.
7. Gently shake the test tube while in the hot-water bath to aid in the dissolving process, if necessary.
8. **BE CAREFUL NOT TO ADD TOO MUCH WATER.** You want to use a minimum amount of water to dissolve the solid. Adding too much water will prevent the crystals from forming.
9. Once all the white powder has dissolved, remove the test tube from the hot-water bath and quickly remove the liquid from any solid impurities that remain in the test tube using a clean plastic pipette. Transfer the liquid from the plastic pipette to a clean, dry test tube.
   - Solid impurities are substances such as sand, dust, or dirt that may have gotten into the sample when it was cleaned up from the floor.
10. If crystals start to form in the test tube in step 9, return the test tube to the hot-water bath until all the crystals dissolve and then remove it from the hot-water bath.

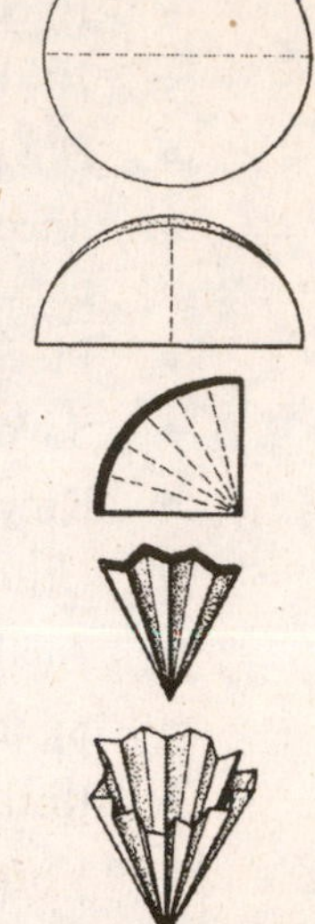

Figure 17.1 Fluting filter paper

11. Allow the liquid in the test tube cool to room temperature. As the solutions cools, crystals should become visible.
12. Prepare a piece of fluted filter paper, as shown in Figure 17.1.
13. Set up a gravity filtration apparatus, as shown in Figure 17.2.
14. Attach the filter flask in Figure 17.2 to a ring stand using a utility clamp, place the filter paper in the funnel and place the funnel on top of the flask.
15. Quickly pour the crystals from the test tube into the filter paper, and allow as much liquid as possible to pass through the filter paper.
16. You can use a little of deionized water to help in the transfer process.

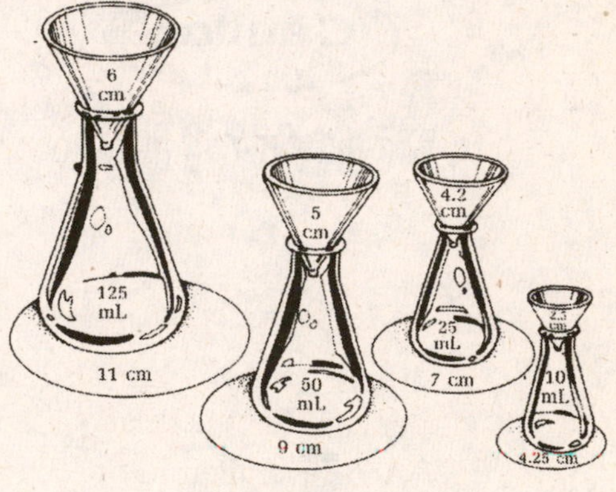

Figure 17.2 Gravity filtration apparatus

17. Remove the filter paper and crystals from the funnel.
18. Using a metal scoopula transfer the crystals to a piece of dry filter paper. Put another piece of filter paper on top of the crystals (making a filter paper– crystal sandwich), and press down on the top piece of filter paper. This will remove any water that remains.
19. Weigh a medium-sized watch glass and record this value on your Report Sheet.
20. Transfer the crystals to the pre-weighed watch glass.
21. Re-weigh the watch glass and the crystals and record this value on your Report Sheet.
22. Set crystals aside until Procedure 9.

Refer to Procedure 1 section of your Report Sheet on p. 184.

PROCEDURE 2

Physical Appearance of White Powders:

1. Obtain the white powder you and your partner are assigned to analyze from your supervisor, and record the sample number on your Report Sheet.
2. Spread a small amount of the white powder on a piece of colored paper to enhance its appearance.
3. Look at the white powder under a hand lens, and record your observations in the Data Table on your Report Sheet.
4. **DO NOT TASTE THE POWDERS.**

Refer to Procedure 2 Data Table of your Report Sheet on p. 185.

PROCEDURE 3

Acetic Acid Test:

1. Obtain a spot plate like the one in Figure 17.3. Each compartment of the spot plate is called a **well**.
2. Place the spot plate on top of a blank sheet of white paper so that all observations may be easily seen.
3. If your spot plate does not contain numbers and letters to identify the wells, label the paper underneath the plate similar to Figure 17.3

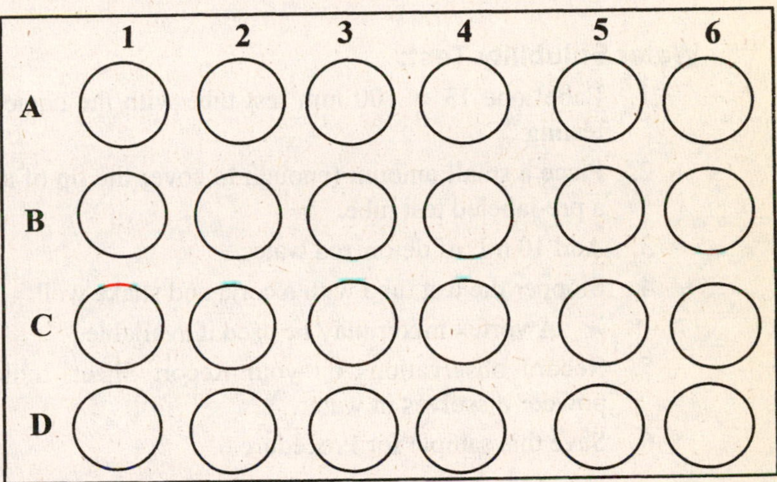

Figure 17.3 Spot plate

   ➢ This labeling system will be used throughout the experiment and will enable you to keep the samples from getting mixed up.

4. Place a small dot of your assigned white powders into two different spot plate wells (the amount should be about the size of this dot •).
5. Add several drops of acetic acid (vinegar) to the well.

   ⚠ If you spill any acetic acid, alert your supervisor immediately.

   ➢ Be careful not to pick up the powder or to dip your dropper into the solution because that will cause contamination.

6. Record your observations on your Report Sheet. Indicate whether, or how completely, the powder dissolves in the acetic acid.
7. Wash off your spot plate in the sink when finished. Dry it with a paper towel for the next test.

Refer to Procedure 3–8 Data Table of your Report Sheet on pp. 185–186.

PROCEDURE 4

Iodine Test:

1. Place a small dot of the first powder into a clean spot plate well.
2. Add several drops of iodine solution to the powder.
   - ➢ Be careful not to pick up crystals or dip your dropper into the solution.
3. Record observations on your Report Sheet. Indicate whether or how completely the powder dissolves in the iodine solution.
4. Wash off your spot plate in the sink when finished. Dry it with a paper towel for the next test.

Refer to Procedure 3–8 Data Table of your Report Sheet on pp. 185–186.

PROCEDURE 5

Water Solubility Test:

1. Label one 13 × 100 mm test tube with the name of the known white powders you are testing.
2. Place a small amount (enough to cover the tip of a microspatula) of the first powder into a pre-labeled test tube.
3. Add 10 mL of deionized water.
4. Stopper the test tube with a cork, and shake well.
   - ➢ A vortex mixer may be used if available.
5. Record observations on your Report Sheet. Indicate whether or how completely the powder dissolves in water.
6. Save this sample for Procedure 6.

Refer to Procedure 3–8 Data Table of your Report Sheet on pp. 185–186.

PROCEDURE 6

Sodium Carbonate Test:

1. To the labeled test tube from Procedure 5, add 5 mL of 1M sodium carbonate solution.
2. Record observations of what happens during the reaction on your Report Sheet. Indicate whether, or how completely, the powder dissolves in sodium carbonate.
3. Watch the reaction closely for the formation of bubbles.
4. Empty the test tube into the sink, and rinse the test tube with deionized water.

Refer to Procedure 3–8 Data Table of your Report Sheet on pp. 185–186.

PROCEDURE 7

Heating Test:

1. Turn on a hot plate to a medium setting (4–5).
2. Label an aluminum-foil dish with the name of the known white powder you are testing.
3. Place a small amount of the white powder into a pre-labeled aluminum-foil dish.

4. Place the aluminum foil dish containing the white powder onto to the hot plate.
5. Allow the powders to heat for several minutes.
6. Record observations on your Report Sheet. Indicate whether or how completely the powder melts or change colors.
7. Remove the aluminum-foil dish from the hot plate using metal tongs.
8. When cool, rinse the aluminum-foil dish with water and let dry.

Refer to Procedure 3–8 Data Table of your Report Sheet on pp. 185–186.

PROCEDURE 8

Isopropyl Alcohol Test:

1. Place a small dot of the known white powder into a spot well.
2. Add isopropyl alcohol dropwise to the well containing the white powder until the well is almost full. Be careful not to pick up crystals or dip your dropper into the solution.
3. Record your observations on your Report Sheet. Indicate whether or how completely the powder dissolves in isopropyl alcohol.
4. Wash off your spot plate in the sink when finished. Dry it with a paper towel for the next test.

Refer to Procedures 3–8 Data Table of your Report Sheet on pp. 185–186.

Fill in the rest of your Data Table with the data collected by your colleagues. Once your Data Table is complete, you are now ready to use the results of the tests completed above to identify Evidence Samples 1, 2, and 3. You will run the same tests on the unknown powders. Comparing the results of these tests with results you obtained above will allow you to identify the unknown substances.

PROCEDURE 9

Identification of Unknown White Powders:

1. If directed, obtain one of the labeled evidence samples.
2. Record the information from the label on the evidence sample on your Report Sheet.
3. Follow Procedures 2–8 on your assigned evidence sample.
4. Record your observations on your Report Sheet. Indicate whether or how completely the powder dissolves or reacts with the test reagent.
5. Compare the test results you got for each of the known powders with the test results for each of the evidence samples.
6. Record the likely identities of Evidence Samples 1, 2, and 3 on your Report Sheet.

Refer to Procedure 9 Data Table of your Report Sheet on p. 187.

NAME (print) ______________________________ DATE ______________
LAST FIRST

LABORATORY SECTION ______ PARTNER(S) ______________________________

REPORT SHEET

Pre-Lab Assignment

Pre-Lab Questions

Find chemical formulas for each of the eight white powders named below. Be sure to include a complete reference.

Baking soda: ______________ Cornstarch: ______________

Sugar: ______________ Salt: ______________

Calcium chloride: ______________ Plaster of paris: ______________

Benzoic acid: ______________ Boric acid: ______________

Reference: ______________________________

PROCEDURE 1

Purify Spilled White Powder:

Initial mass of Evidence Sample 3 ______________ g

Mass of watch glass: ______________ g

Mass of watch glass and crystals ______________ g

Recovered mass of sample ______________ g

Indicate which known white powders and/or evidence samples you have been asked to analyze. You will be given the test results for the other known white powders and evidence samples from your colleagues.

Known white powder: ______________________________

Known white powder: ______________________________

Evidence Sample number: ______________________________

NAME (print) ____ DATE ____________________
LAST FIRST

LABORATORY SECTION ______ PARTNER(S) ____

PROCEDURE 2

Physical Appearance of Known White Powders:

Look at each powder under a hand lens. Record observations for all eight powders.

White Powder	Physical Appearance
Baking powder	
Cornstarch	
Sugar	
Salt	
Benzoic acid	
Boric acid	
Calcium chloride	
Plaster of paris	

PROCEDURES 3–8

Known Samples:

Circle the white powder you analyzed. Record your observations for the white powder you analyzed for Procedures 3–8 in the Data Table below. Fill in the rest of the table with data from your colleagues. The Data Table can be found on page 186 of this Lab Manual.

NAME (print) ______________________________ DATE ______________
LAST FIRST

LABORATORY SECTION _______ PARTNER(S) ______________________________

Known Samples:

	Procedure 3	Procedure 4	Procedure 5	Procedure 6	Procedure 7	Procedure 8
White Powder	**Acetic Acid Test**	**Iodine Test**	**Solubility Test**	**Sodium Carbonate Test**	**Heating Test**	**Isopropyl Alcohol Test**
Baking powder						
Cornstarch						
Sugar						
Salt						
Benzoic acid						
Boric acid						
Calcium chloride						
Plaster of paris						

NAME (print) ______________________________ DATE ____________
LAST FIRST

LABORATORY SECTION ______ PARTNER(S) ______________________________

PROCEDURE 9

Evidence Samples: Physical Appearance:

Look at each powder under a hand lens. Record observations for all evidence samples.

Evidence Sample	Physical Appearance
Evidence Sample 1	
Evidence Sample 2	
Evidence Sample 3	

Evidence Samples:

Record your observations for Procedures 3–9 in the Data Table below.

	Procedure 3	Procedure 4	Procedure 5	Procedure 6	Procedure 7	Procedure 8
White Powder	Acetic Acid Test	Iodine Test	Solubility Test	Sodium Carbonate Test	Heating Test	Isopropyl Alcohol Test
Evidence Sample 1						
Evidence Sample 2						
Purified Evidence Sample 3						

NAME (print) ______________________________________ DATE ____________________
LAST FIRST

LABORATORY SECTION ______ PARTNER(S) ________________________________

DATA ANALYSIS

Based on your observations on your Report Sheet, identify the white powders.

Evidence Sample 1, powder found in girls' bathroom, is______________________________

Evidence Sample 2, powder found in boys' locker room, is____________________________

Evidence Sample 3, powder found on student, is_____________________________________

SOURCES OF ERROR

Once the data have been collected, the possibility of error in the results must be addressed. Please cite at least two random sources of error and briefly explain how each source of error may have affected your data. For a review of error analysis, see *Error Analysis in the Chemistry Laboratory* in the Appendix at the end of this manual.

POST-LAB QUESTIONS

1. Write the balanced chemical reaction that occurred for the reaction of calcium chloride and sodium carbonate.

2. Explain how the result you observed for this reaction is supported by the balanced reaction.

a. Recall from the experiment which test produced bubbling. Explain this observation by writing a balanced chemical equation.

NAME (print) ______________________________________ DATE ______________________
LAST FIRST

LABORATORY SECTION ______PARTNER(S) ____

b. Based on your qualitative test of white powders, put together a field kit for Officer Look to use on calls in the future where a mysterious white powder is involved. Explain what will be in the kit and why.

CONCLUSION

Describe what you and your fellow analysts have determined about the white powders found at the high school. State your conclusions and the reasoning behind them in complete sentences. Be sure that your conclusion is complete and concise.

Experiment 18

Detection of Gunshot Residues (GSRs)

PRE-LAB ASSIGNMENT

Reading:

1. Read the experiment.
2. Refer to Chapter 16 of *Adventures in Chemistry*.

Question:

3. The concentration of a stock solution of gunshot residue is 2.00 M. You pipette 2.50 mL of this stock solution into a 10.0 mL volumetric flask and bring the volume of the volumetric flask up to 10.0-mL with deionized water. What is the concentration of the new solution you have just created?

INTRODUCTION

A night of random shooting has left many residents waking up to find their car windshield shot out. Given the high number of reports, the local authorities are unable to visit each crime scene to collect evidence and talk to potential witnesses. To help with the investigation, members of the Firearms Identification Division (FID) at UniChem International have been called in to assist. According to one witness, two young men were seen leaving the scene of one of the vandalized cars on bicycles. The two young men were found a short time later and were questioned by FID agents. The agents ask permission to collect evidence from both suspects that may be used to determine whether either of the young men recently fired a weapon.

If a suspect has fired a weapon recently, a wet swab of the suspect's left and/or right hand will yield an absorbance greater than 1 of **gunshot residue** *(GSR)* in the spectrophotometer. Twelve evidence swabs were collected by FID agents and were labeled as follows:

Suspect A:	Swab Location	Suspect B:	Swab Location
Evidence Swab 1:	Back of right hand	**Evidence Swab 7:**	Back of right hand
Evidence Swab 2:	Right-hand thumb web	**Evidence Swab 8:**	Right-hand thumb web
Evidence Swab 3:	Right-hand palm	**Evidence Swab 9:**	Right-hand palm
Evidence Swab 4:	Back of left hand	**Evidence Swab 10:**	Back of left hand
Evidence Swab 5:	Left-hand thumb web	**Evidence Swab 11:**	Left-hand thumb web
Evidence Swab 6:	Left-hand palm	**Evidence Swab 12:**	Left-hand palm

An unused swab also was sent to be used as control sample. Working in groups of three, your job is to perform GSR determination for the evidence samples collected to determine if either suspect fired a gun recently, and if so, to determine which hand did the firing.

If a firearm is discharged, the individual who discharged the firearm can be identified by GSR determination. One method is to analyze trace metals arising from the gun powder and primer residues that are deposited on the hands of the shooter using a spectrophotometer. The firing pin of the weapon is released when the suspect pulls the trigger. The firing pin's job is to strike the primer, which, in turn, ignites the gun powder. As the powder burns, gases build up in the chamber forcing the bullet to propel through the barrel and causing gun powder and primer to propel toward the shooter. Most primers, with the exception of most .22-caliber bullets contain different combinations of lead styphnate ($PbC_6H_3N_3O_8$), barium nitrate ($Ba(NO_3)_2$), and antimony sulfide (SbS). Residues of these compounds may be deposited on the thumb web and back of the firing hand of someone who has fired a weapon recently. GSR also may be present on the palm of the person who has come in contact with a recently fired weapon. The metals analyzed in a GSR determination are commonly antimony (Sb) from SbS and barium (Ba) from $Ba(NO_3)_2$. The lead present in lead styphnate is least useful to FID agents owing to its naturally occurring abundance in the environment. If barium and/or antimony are found above established threshold levels, it is considered substantial evidence for the presence of a GSR in a firearm-discharge situation.

For our purposes today, we will analyze for the presence of GSR. The first part of the analysis requires making a standard curve to which all evidence samples can be compared. You will begin by making a series of four standard solutions from a 4×10^{-4} M GSR stock solution (4×10^{-4} M expresses the concentration of the stock solution).

MORE ABOUT CONCENTRATIONS: The large **M** is a unit representing **molarity**. It is equivalent to the number of moles per liter. For example, if you purchase a chemical solution that is 10.0 M HCl, this means that you have 10.0 mol HCl per 1 liter of water. If you start with the 10.0 M HCl as your stock solution, you can make a new solution of a smaller concentration from it. How? The molarity of your starting solution (stock, mol/L) times the amount of starting solution you use (L) **is equal to** the molarity of the new solution (mol/L) times the amount of new solution you make (L). This is sometimes represented as

$$M_1V_1 = M_2V_2$$

where M_1 equals the molarity of the stock solution, V_1 equals the volume of the stock solution, M_2 equals the molarity of the diluted solution, and V_2 equals the volume of the diluted solution. If three of the variables are known, you can solve for the fourth variable.

For example: Brittney buys a 1-L bottle of 5.0 M acetic acid. For her photographic fixer she needs 20.0 L of a 0.1 M acetic acid solution. To make her new solution, how many liters of the 5.0 M acetic acid would she need to use?

Here is how to set up the problem:
Remember $M_1V_1 = M_2V_2$ where
$M_1 = 5.0$ *M* acetic acid
$V_1 = ?$ L
$M_2 = 0.1$ *M* acetic acid
$V_2 = 20.0$ L

or (5.0 mol/L) × (? L) = (0.1 mol/L) × (20.0L).
Solving for V_1 gives 0.4 L, or 400 mL

Therefore, Brittney would need to pour 400 mL of her purchased acetic acid into her container and bring the volume of water up to 20.0 L.

After making the four standard solutions, you will measure the absorbance of each using a spectrophotometer. The basic principle behind the analysis is that the more particles there are in a solution, the higher is the absorbance of light detected by the spectrophotometer. A **spectrophotometer** is an instrument that compares the amount of light transmitted by the blank sample with the amount of light transmitted by your sample. The more concentrated (darker color) the sample is, the more light it will absorb compared with the blank. The amount of light absorbed at a particular wavelength of a sample is known as the **absorbance**. If absorbance values are obtained for a series of solutions with different known concentrations, a **standard curve**, or **calibration curve**, can be obtained. The standard curve then can be used to determine the concentrations of unknown samples. This is accomplished by measuring the absorbance of the unknown samples at the same wavelength used to generate the standard curve.

LEARNING OBJECTIVES

Be able to:

- Understand the concept of molarity and solution dilution.
- Create and use a standard curve.
- Use and operate a spectrophotometer.

APPARATUS

Chemicals:

- Standard stock solution of GSR, 4×10^{-4} M
- 12 evidence samples
- Blank sample
- Deionized water bottles

Equipment:

- 4 10.0-mL volumetric flasks with stoppers
- Roll of colored tape used for labeling
- 10 plastic disposable pipettes
- 11 plastic disposable cuvettes
- Cuvette rack
- 500μL automatic pipette
- Box of Kim wipes
- Spectrophotometer

- Are your safety glasses on?
- Rinse all cuvettes with ethanol in a fume hood.
- Collect the ethanol in a labeled flammable container.
- Ethanol is flammable. **Do Not** use open flames.

Clean-up: 

- All liquids can go down the drain.
- Clean cuvettes as instructed by your supervisor.
- Wash your hands thoroughly with soap and water before you leave the laboratory.

PROCEDURE 1

Preparing Standard Solutions:

1. Using tape, label four 10.0-mL volumetric flasks as follows: Standard 1, Standard 2, Standard 3, and Standard 4.
   - ➢ **DO NOT** cover the meniscus with the label.
2. Using the automatic pipette, pipette the 4×10^{-4} M gunshot residue stock solution into the pre-labeled volumetric flasks as follows:
   - Standard 1: Add 200 μL
   - Standard 2: Add 300 μL
   - Standard 3: Add 400 μL
   - Standard 4: Add 500 μL
3. Record the amount of GSR stock solution added on your Report Sheet.
4. Carefully bring the volume of liquid up to the etched white ring level (bottom of meniscus should sit on the line) with deionized water for the flask labeled "Standard 1." You will find a plastic dropper a very useful tool for doing this.
5. Place stoppers into top of the Standard 1 flask, hold firmly, and turn the flask upside down at least 10 times to ensure sufficient mixing.
6. Repeat steps 4– 6 for the remaining standard solutions.
7. Set up a cuvette rack of five cuvettes. Label the cuvette rack with tape (*Note:* You cannot mark the cuvettes) to identify the blank and the Standards 1, 2, 3, and 4.
8. Rinse and then fill the blank cuvette with deionized water until it is approximately three-quarters full.
9. Rinse and then fill the Standards 1, 2, 3, and 4 cuvettes with the respective standard solutions (approximately three-quarters full).
   - ➢ You are now ready to start using the spectrophotometer.

Refer to Procedure 2 section of your Report Sheet on p. 198.

PROCEDURE 2

Important Spectrophotometer Notes:

- Follow your supervisor's instructions for the proper use of the spectrophotometer.
- Wipe the outside walls of **EACH** cuvette with a Kim wipe to remove all smudges, dust, and liquid before inserting into the cuvette holder.
- Hold the cuvette by the two frosted sides.
- Make sure that the clear sides of the cuvettes are in the pathway of the light beam when placed in the spectrophotometer cuvette holder.
- Be sure to run a blank before collecting data on the evidence samples.
- You are now ready to acquire absorbance data using your standard solutions.

Collecting Absorbance Data for Standard Solutions:

1. Replace the blank in the cuvette holder with Standard Solution 4.
2. Following the instructions for your spectrophotometer, determine the wavelength at maximum absorbance. Record the wavelength on your Report Sheet.
   - ➢ From this point on, all absorbance readings will be made at this specific wavelength! This means, **DO NOT CHANGE THE WAVELENGTH WHILE COLLECTING DATA.**
3. Record the absorbance value on your Report Sheet for Standard solution 4.
4. Replace the cuvette in the cuvette holder with Standard Solution 3.
5. Repeat steps 5–6 for Standard Solution 3.
6. Continue to read the remaining two standard solutions by replacing the cuvette with another sample and reading its absorbance at your specific wavelength.
7. Record the absorbance readings for Standard Solution 2 and Standard Solution 1 on your Report Sheet.
8. What other quality-control samples might you consider analyzing?

Refer to Procedure 2 section of your Report Sheet on p. 198.

PROCEDURE 3

Graphing:

1. After all the standards samples are analyzed, you will create a standard curve by graphing the known concentrations of your set of standards relative to the respective absorbance readings.
2. You will first need to calculate the concentrations of Standards 1, 2, 3 and 4 and record these values on your Report Sheet. The calculation for determining concentrations of standards is the same calculation as in the Pre-Lab Assignment.
3. Construct a graph using Excel®, with the x values (the independent variable), which are the concentrations of the diluted solutions, listed in the first column. The y values (the dependent variable), which are the absorbance readings, need to be listed in the second column.
4. To make the graph, click on the graphing icon in the uppermost tool bar.
5. Choose **x-y Scatter** as the graph type.
6. Now click **Next** to be able to label the axes (don't forget the units) and title.

7. When your graph is completed click on **Finish**.
8. You now click on **Chart**, and select **Add a Trend Line.**
9. Click the **Options** tab, and then click on the box next to **Display Equation on Chart**.
10. You also will want to check off the box next to **Display R-squared Value on Chart**.
11. Print your graph.

Attach Graph to your Report Sheet on p. 202.

You are now ready to collect absorbance readings on one of the suspects accused of shooting out the windshields. On completion the data collection on your suspect, your data will be exchanged with the data of another group of FID agents who analyzed the evidence on the other suspect. Using the standard curve generated in Procedure 3, you will be able to determine if one of the suspects is guilty and which hand was used to fire the gun.

PROCEDURE 4

Collecting Absorbance Data for Evidence Samples:

1. Obtain evidence samples from one of the suspects. Record the suspect on your Report Sheet.
2. Setup a cuvette rack of six cuvettes. Label the cuvette rack with tape (*Note:* You can not mark the cuvettes) to identify the samples as follows: 1: back of right hand, 2: right-hand thumb web, 3: right-hand palm, 4: back of left hand, 5: left-hand thumb web, and 6: left-hand palm, which correspond to Suspect A.
3. Rinse and then fill the cuvettes with the respective evidence samples. You are now ready to collect absorbance readings for the six evidence samples.
4. Place the cuvette containing Evidence Sample 1 into the cuvette holder.
5. Record the absorbance value on your Report Sheet for Evidence Sample 1.
6. Replace the cuvette in the cuvette holder with Evidence Sample 2.
7. Repeat step 8 for Evidence Sample 2.
8. Continue to collect absorbance data for the four remaining evidence samples by replacing the cuvette with another sample and reading its absorbance at your specific wavelength.
9. Record the absorbance readings on your Report Sheet.
10. Record your data on the group spreadsheet as instructed by your supervisor.

Refer to Procedure 4 section of your Report Sheet on p. 199.

DATA ANALYSIS

1. You will use the equation of the line of your standard curve to determine the concentration of GSR in each of the evidence samples tested.
2. The general equation for a line is $y = mx + b$, where y is the y intercept, m is the slope of the line, and x is the x intercept.
3. To determine the concentrations of the evidence samples, use your equation of the line from your standard curve. Substitute the absorbance value for Evidence Sample 1 for y in the equation of the line and solve for x. The value of x is equal to the concentration of Evidence Sample 1.
4. Record the concentration for Evidence Sample 1 on your Report Sheet.

5. Repeat steps 4–6 for the remaining five evidence samples.
6. Record the concentrations of the GSR determinations on your Report Sheet.
7. Exchange your data from Procedure 4 with the data of other FID agents of UniChem International who analyzed the evidence samples from the **other** suspect.
8. Record the names of the agents you exchanged data with on your Report Sheet.
9. Record the absorbance values and concentrations on your Report Sheet.
10. Identify whether each evidence sample tested positive or negative for GSR.
11. Remember that if an evidence sample has an absorbance greater than 1GSR, it has tested positive for gunshot residue.

Refer to Procedure 4 section of your Report Sheet on p. 199.

NAME (print) ______________________________ DATE ______________
LAST FIRST

LABORATORY SECTION ____ PARTNER(S) ______________________________

REPORT SHEET

Pre-Lab Assignment

Pre-Lab Question:

The concentration of a stock solution of gunshot residue is 2.00 M. You pipette 2.50 mL of this stock solution into a 10.0-mL volumetric flask and bring the volume of the volumetric flask up to 10.0 mL with deionized water. What is the concentration of the new solution you have just created? Show all your work.

PROCEDURE 2

Absorbance Readings of Standard Solutions:

Spectrophotometer number: ______________

Wavelength: ____________________ nm

Standard Solution	Amount of GSR (μL)	Absorbance Reading	Concentration
1			
2			
3			
4			
Blank			

NAME (print) ____ DATE ____________________
LAST FIRST

LABORATORY SECTION ______PARTNER(S) ____

Sample Calculation Concentration Determination: Show how the concentration of Standard Solution 1 was determined.

PROCEDURE 4

Absorbance Readings for Evidence Samples

Fill in the following Data Table. Be sure to get all the data from the group spreadsheet.

Evidence Sample Number	Analyst	Absorbance Reading	Concentration (M)
1			
2			
3			
4			
5			
6			
7			
8			
9			
10			
11			
12			

NAME (print) ____ DATE ______________________________
LAST FIRST

LABORATORY SECTION ______ PARTNER(S) ____

PROCEDURE 3

Graphing of Standard Curve:

Equation of the line obtained from standard curve ____________________

Sample Calculation: Using the equation of the line from your standard curve, show how the concentrations of one of the evidence samples were determined.

SOURCES OF ERROR

Once the data have been collected, the possibility of error in the results must be addressed. Please cite at least three random sources of error and briefly explain how each source of error may have affected your data. For a review of error analysis, see *Error Analysis in the Chemistry Laboratory* in the Appendix at the end of this manual.

NAME (print) ______________________________________ DATE ____________________
LAST FIRST

LABORATORY SECTION ______ PARTNER(S) ______________________________________

POST-LAB QUESTIONS

1. The state laboratory analytical license is due for recertification, and you are part of the inspection team. You inspect the notebook of one of the analysts in the Firearms Division. Regardless of what page you look at, what types of samples should every analyst run or include in their collection of data to ensure quality control? Briefly explain.

2. During your setup one of the 10-mL volumetric flasks breaks. The only other volumetric flasks in the lab are 250 mL. In order to keep the concentrations the same (as you had today) for your standard curve, how much of the stock solution will you have to pipette into Standard 1–4, respectively? Please show all work.

3. You are up for a promotion and are hoping to impress your supervisor with your dazzling efficiency. After running a particularly important evidence swab you realize that the GSR exceeds the highest standard for your standard curve. Name one procedural change that you could quickly make to accommodate the special sample. Explain a second change in the procedure (different from the first) that you also might try.

NAME (print) ____ DATE ____________________
LAST FIRST

LABORATORY SECTION ______ PARTNER(S) ____

CONCLUSION

Describe what you and your fellow Firearms Identification Division analysts have determined about the amount of Gunshot Residue (if any) in the evidence samples. Does any evidence sample indicate an absorbance greater that 1? If yes, which suspect fired the gun? Does any evidence sample(s) indicate a particular hand(s) being used? How can you support this with your data? What other things can you conclude or state? State your conclusions and the reasoning behind them in complete sentences. Be sure that your conclusion is complete and concise.

Attach Graph Here:

Experiment 19

Identification of Six Common Types of Plastics

PRE-LAB ASSIGNMENT

Reading:

1. Read the experiment and refer to Chapter 18 of *Adventures in Chemistry*.

Research:

2. Find two different uses or applications for each of the following types of plastics: polyethylene terephthalate, high-density polyethylene, polyvinyl chloride, low-density polyethylene, polypropylene, and polystyrene. These uses or applications should be different from the ones given in the lab handout.
3. Record the uses or applications on your Report Sheet.
4. Include complete references for where you found the information.

INTRODUCTION

There has been a rash of burglaries in the small towns surrounding the home office of UniChem International. According to the local Sheriff's Department, the missing items included cash and typical quick-cash items that were hauled off in a truck during the middle of the day. Members of the Criminal Investigation Unit (CIU) at UniChem International were called in to assist the local authorities and were delighted by the one and only piece of evidence found at one of the burglarized homes—a plastic swipe card with all identification removed from the card. The modus operendi (MO) of the burglars was to use a plastic card, such as the plastic swipe cards used by motels and hotels, to jimmy the door locks. CIU officers have questioned managers at the local motels that use swipe cards for entrance instead of keys. The type of plastic for each swipe card collected is different for each manager interviewed.

CIU has obtained a plastic swipe card from each motel and has tagged them as evidence. The evidence samples have been labeled as follows:

Known Sample 1: Sleepy Time Motel
Known Sample 2: Tuckered Out Inn
Known Sample 3: Hodge Podge Lodge
Known Sample 4: Settle Down Inn
Known Sample 5: Dew Drop Inn
Known Sample 6: Sweet Dreams Motel
Evidence Sample 7: Plastic card found at scene

Six of 60,000 types of plastics make up 70% of the plastics used in the United States. Typically, the type of plastic a given object is made from is found stamped onto the object as a number in the center of a triangle. The six plastic types are polyethylene terephthalate (#1), high-density polyethylene (#2), polyvinyl chloride (#3), low-density polyethylene (#4), polypropylene (#5), and polystyrene (#6). The industrial coding for these six types of plastics enables recycling and is known nationally as follows:

#1 Polyethylene terephthalate (PETE): The most commonly recycled plastic. Used to make 2- liter soda bottles and liquor bottles. Recycled into products used for nonfood items (e.g., laundry detergent bottles, carpeting, and fleece).

#2 High-density polyethylene (HDPE): Commonly used to make milk and juice bottles. Recycled into products such as pails, trash cans, toys, piping, and the bases of soft drink bottles.

#3 Polyvinyl chloride (PVC): Used to make flooring, shower curtains, siding for buildings, and garden hoses. Currently not recycled.

#4 Low-density polyethylene (LDPE): Used to make cellophane wrap, disposable diaper liners, and squeeze bottles. Not commonly recycled.

#5 Polypropylene (PP): Used to make packaging pipes, tubes, sports wear, long underwear, and tents. Not commonly recycled.

#6 Polystyrene (PS): Also known as Styrofoam, used to make coffee cups, egg boxes, packing peanuts, and take-out food packaging. It is recycled in some areas for insulation, plastic wood, and writing pens.

Preliminary analyses of the Known Samples 1–6 have identified the type of plastic used to prepare each swipe cards, and the results are given below:

Known Sample 1: PETE
Known Sample 2: HDPE
Known Sample 3: PVC
Known Sample 4: LDPE
Known Sample 5: PP
Known Sample 6: PS

Working in pairs, you will develop the **qualitative** method needed to analyze and determine the plastic used to make the plastic swipe card found at the scene of one of the burglaries. Your supervisor will assign each pair of agents one of the evidence samples to analyze. The results of your analyses will be compiled with the results collected by the other members of your division in order to develop the qualitative method, which will be used to identify the plastic used to make unknown Evidence Sample 7. If the Evidence Sample 7 plastic can be identified as one of the known types of plastic, then a connection may be made to a particular motel. This would be an ideal way to locate the thieves, because some of the stolen goods still may be a motel room.

> **Qualitative analysis** in the strict chemical sense has to do with designing a chemical test to allow the identification of the components of any given substance or mixture. In a broad context, this type of analysis typically involves a comparison of characteristics (e.g., color, shape, and consistency) that are not always quantifiable. In other words, the comparisons are not usually measured, merely observed.

LEARNING OBJECTIVES

Be able to:

- Examine some of the properties of these six types of plastics.
- Develop a qualitative classification and identification scheme that will aid in the identification of the plastic swipe card.
- Inform the local authorities as to which lodging establishment the plastic swipe card came from.

APPARATUS

Chemicals:

- 50 % Ethanol-water solution
- 10% Sodium chloride (NaCl) solution
- Deionized water

Equipment:

- 2 pieces of each samples of plastic types #1–#6 cut into strips.
- 2 pieces of Evidence Sample 7 cut into strips
- Roll of colored tape for labeling
- Several pairs of scissors (can be shared)
- Bunsen burner
- Bunsen burner Striker
- Pair crucible tongs or 8-inch metal tweezers
- Blue litmus paper
- 6-in Piece copper wire attached to a cork
- Plastic weighing boats or dishes

- Are your safety glasses on?
- Perform the melt, ignition, and copper wire test in a fume hood.
- Tie back long hair and roll up baggy sleeves before doing any work with Bunsen burners.

Clean-up:

- No special clean up is necessary for this experiment. All liquid can go down the drain.
- Pieces of plastic can go into the trash once cooled to room temperature.
- Wash your hands thoroughly with soap and water before you leave the laboratory.

PROCEDURE 1

NOTE: Evidence Samples 1–7 may contain plastic pieces that may or may not have color. You will not need to design your comparative analysis of plastic types to consider coloring differences.

Density Test:

1. Place a piece of tape on three 250-mL beakers and label them as follows: Beaker 1 density = 0.94 g/mL, Beaker 2 density = 1.0 g/mL, and Beaker 3 density = 1.08 g/mL.
2. Fill Beaker 1 approximately half full with 50% ethanol-water solution.
3. Fill Beaker 2 approximately half full with deionized water.
4. Fill Beaker 3 approximately half full with 10% NaCl solution.
5. Obtain two pieces of Known Sample 1, which contains #1 plastic.
6. Cut each of the two pieces of plastic into three equal parts. Three of the pieces of #1 plastic you just cut will be used to perform the density test.
7. Place the other three pieces of #1 plastic you've just cut into a weighing boat labeled #1 and set aside for use in the next three qualitative tests.
8. Place one piece of #1 plastic into each of the three labeled beakers.
9. Using a glass stirring rod, push the piece of #1 plastic in Beaker 1 under the liquid.
10. Observe what the piece of plastic does once you've removed the stirring rod.
    - If the plastic floats, the plastic has a density less than the given test solution.
    - If the plastic sinks, the plastic has a density greater than the given test solution.
11. Repeat the procedure for Beakers 2 and Beaker 3.
12. Record all observations on your Report Sheet.
13. Remove #1 plastic from all three beakers.
14. Repeat the density test on each of the other plastics #2–#6 and the Evidence Sample 7.
15. Record your observations for plastics #2–#6 and the Evidence Sample 7 on your Report Sheet.

Refer to Procedure 1 section of your Report Sheet on p. 209.

PROCEDURE 2

Melt Test:

1. Perform the melt test in a fume hood.
2. Using a piece of #1 plastic from the weighing boat from Procedure 1, place the plastic on the tip a metal scoopula.
3. Light a Bunsen burner, and adjust the flame until a blue cone can be seen in the middle of the flame.
4. Carefully place the tip of the scoopula containing the piece of plastic into the blue part of a Bunsen burner flame.
5. Heat slowly by moving the tip of the scoopula in and out of the flame so that the plastic does not catch on fire, and record your observations on your Report Sheet as the plastic melts.
6. Once the plastic piece has melted completely, remove it from the flame.
7. Allow the sample to cool, and note the appearance of the plastic.

8. Test the flexibility of the plastic by trying to bend it. Record your observations on your Report Sheet.
9. Repeat the melt test on each of the other plastics #2–#6 and the Evidence Sample 7.
10. Record your observations for plastics #2–#6 and the Evidence Sample 7 on your Report Sheet.

Refer to Procedure 2 section of your Report Sheet on p. 210.

PROCEDURE 3

Ignition Test:

1. Perform the ignition test in a fume hood.
2. Obtain a piece of #1 plastic from the weighing boat, and hold one end of it with a pair of tongs (or tweezers).
3. Light a Bunsen burner, and adjust the flame until a blue cone can be seen in the middle of the flame.
4. Place the plastic directly in the flame.
5. Record your observations on your Report Sheet with respect to what happened to the plastic. Note if the plastic burns, note the color of the flame, and note whether or not smoke is visible. (Record color of smoke, if present.)
6. Observe what happens to the plastic when you remove it from the flame?
7. Record this observation on your Report Sheet.
8. Determine if the smoke is acidic by holding a piece of wet blue litmus paper in the smoke.
9. If the paper turns red, the smoke is **ACIDIC**. If there is no color change, the smoke is not acidic.
10. Record your observations for the acidity of the smoke of plastic #1 on your Report Sheet.
11. When testing on the #1 plastic is complete, put the burning plastic in a 400-mL beaker of water to put out the flame.
12. Repeat the ignition test on each of the other plastics #2–#6 and the Evidence Sample 7.
13. Record your observations for plastics #2–#6 and the Evidence Sample 7 on your Report Sheet.

Refer to Procedure 3 section of your Report Sheet on p. 210.

PROCEDURE 4

Copper Wire Test:

1. Perform the copper wire test in a fume hood.
2. Remove a piece of #1 plastic from the weighing boat and place it on a watch glass.
3. Obtain a piece of 6-in. copper wire that is attached to a cork.
4. Light a Bunsen burner, and adjust the flame until a blue cone can be seen in the middle of the flame.
5. Place the tip of the copper wire directly into the flame.
6. Heat the copper wire until the green color is not longer visible.
7. While the tip of the copper wire is still hot, pierce the piece of #1 plastic that is on the watch glass.
8. The piece of plastic should stick to the tip of the copper wire.

9. Place the tip of the copper wire and the piece of plastic into the blue part of the flame and observe the color of the flame.
10. Record the color of the flame for the #1 plastic on your Report Sheet.
    - If you observe a flash of green color in the flame or if the flame turns greens, this indicates that the halogen chlorine is present in the plastic.
11. When testing on the #1 plastic is complete, put the burning plastic in a 400-mL beaker of water to put out the flame.
12. Repeat the copper wire test on each of the other plastics #2– #6 and the Evidence Sample 7.
13. Record your observations for plastics #2–#6 and the Evidence Sample 7 on your Report Sheet.

Refer to Procedure 4 section of your Report Sheet on p. 211.

You are now ready to use the results of the tests just completed to identify Evidence Sample 7. Comparing the results of the tests you ran on the known plastic samples with the results you obtained on the unknown Evidence Sample 7 will allow you to identify the unknown plastic used in the swipe card found at the scene. In addition to this, you also will be able to develop an identification scheme for identifying plastic types that can be used for future evidence samples.

NAME (print) ______________________________ DATE ______________
LAST FIRST

LABORATORY SECTION ______ PARTNER(S) ______________________________

REPORT SHEET

Pre-Lab Assignment

Pre-Lab Question:
Find two different uses or applications for each of the following plastics: polyethylene terephthalate, high-density polyethylene, polyvinyl chloride, low-density polyethylene, polypropylene, and polystyrene. These should be different from the ones given in the lab handout. Be sure to include a complete reference.

#1 PETE:

#2 HDPE:

#3 PVC:

#4 LDPE:

#5 PP:

6 PS:

Reference(s): ______________________________

PROCEDURE 1

Density Test: For the different types of plastics tested, indicate whether the plastic "sinks" or "floats" in the different solutions. If your observations were not clear write "inconclusive".

Type of Plastic	50% Ethanol-Water Density = 0.94 g/mL	Deionized Water Density = 1.00 g/mL	10% NaCl Density = 1.08 g/mL
#1 PETE			
#2 HDPE			
#3 PVC			
#4 LDPE			
#5 PP			
#6 PS			
Evidence Sample 7			

NAME (print) ____ DATE ____________________
LAST FIRST

LABORATORY SECTION ______ PARTNER(S) ____

PROCEDURE 2

Melting Test: For the different types of plastics tested, record what you observed as the plastic melted and the appearance and flexibility after it cooled.

Type of Plastic	Observations	Appearance and Flexibility
#1 PETE		
#2 HDPE		
#3 PVC		
#4 LDPE		
#5 PP		
#6 PS		
Evidence Sample 7		

PROCEDURE 3

Ignition Test: For the different types of plastics tested, record what you observed as the plastic ignited and burned. Did you observe smoke? If yes, was the smoke acidic?

Type of Plastic	Observation	Smoke	Acidic (Yes or No)
#1 PETE			
#2 HDPE			
#3 PVC			
#4 LDPE			
#5 PP			
#6 PS			
Evidence Sample 7			

NAME (print) ______________________________________ DATE ____________________
LAST FIRST

LABORATORY SECTION ______ PARTNER(S) ________________________________

PROCEDURE 4

Copper Wire Test: For the different types of plastics tested, record what you observed when the copper wire and plastic sample are returned to the flame. Did you observe a green flame?

Type of Plastic	Observation	Color of Flame?
#1 PETE		
#2 HDPE		
#3 PVC		
#4 LDPE		
#5 PP		
#6 PS		
Evidence Sample 7		

DATA ANALYSIS

Evidence Sample Identification:

1. Based on your observations on your Report Sheet, devise a qualitative scheme that will help you to identify the unknown plastic from Evidence Sample 7 that was used in the burglaries.

NAME (print) ______________________________________ DATE ______________________
LAST FIRST

LABORATORY SECTION ______ PARTNER(S) ______________________________________

2. Once you have identified the plastic used in the crimes, you can direct the authorities to the motel from which the plastic swipe card came. The names of the motels and the type of plastic their swipe cards are made from are listed below:

Plastic #1: Sleepy Time Motel
Plastic #2: Tuckered Out Inn
Plastic #3: Hodge Podge Lodge
Plastic #4: Settle Down Inn
Plastic #5: Dew Drop Inn
Plastic #6: Sweet Dreams Motel

Identification of Evidence Sample 7: ______________________________

Motel using plastic type: ______________________________

SOURCES OF ERROR

Once the data have been collected, the possibility of error in the results must be addressed. Please cite at least two random sources of error and briefly explain how each source of error may have affected your data. For a review of error analysis, see *Error Analysis in the Chemistry Laboratory* in the Appendix at the end of this manual.

NAME (print) ____ DATE ______________________
LAST FIRST

LABORATORY SECTION ______ PARTNER(S) ____

POST-LAB QUESTIONS

1. You have just been put in charge of recycling for an environmentally friendly city. Currently, all your plastics are being sorted by hand. An increased demand for shredded PET containers prompts you to consider a more scientific means of sorting the types of plastic. Based on your experience with the various methods for testing plastic types propose a more efficient way of recycling plastic based on sorting out the PET containers.

2. Below is a list from the UniChem International Forensic Investigators Identification Library of cars made with the known types of plastics.

   Plastic #1: Honda
   Plastic #2: Toyota
   Plastic #3: Ford
   Plastic #4: Mazda
   Plastic #5: Subaru
   Plastic #6: Chevy

   Use your newly created identification scheme as a reference to answer the following scenario:

   You are off on a tropical island vacation when a "hot" evidence sample of plastic car fragments comes in for identification. Your supervisor picks up your lab book and follows your procedure to identify the sample. Following your scheme she determines the sample to be from a Chevy. Over the phone, while sipping your icy tropical fruit drink you confirm what she should have observed for each of the four tests. Give your confirmation summary.

NAME (print) ______________________________ DATE ______________________
LAST FIRST

LABORATORY SECTION ______ PARTNER(S) ______________________________

CONCULSIONS

Describe what you and your fellow Criminal Investigation Unit analysts have determined about the plastic used to make the plastic swipe card found at the crime scene. Which establishment uses this type of plastic for its swipe cards? State the recommendations that you will make to the Sheriff's department and the reasoning behind them in complete sentences. Be sure that your conclusion is complete and concise.

Experiment 20

Making Polymers

PRE-LAB ASSIGNMENT

Reading:

1. Read the experiment.

Research:

2. Research the chemistry of polymers (see Chapter 18 of *Adventures in Chemistry*).

Question:

3. Answer the Pre-Lab Questions.

INTRODUCTION

Polymers are large molecules that are created by linking together repeating units of smaller molecules, known as **monomers**. These monomers bond together into long chains, which, in turn, can cross-link with each other. Typical polymer molecules contain many thousands or even millions of atoms. Polymers are extremely important in many aspects of our lives, from storage and transmission of our genetic material (DNA and RNA) to making plastic containers (polystyrene and polypropylene), clothing (Orlon and Acrilan) and non-stick pots and pans (Teflon).

There are two basic types of polymers: synthetic polymers, which are created by scientists and biopolymers, which are produced by living organisms. Synthetic polymers include a great variety of materials that we use today, such as polyester, polyvinyl chloride (PVC), Teflon, and Kevlar. Biopolymers include starch and cellulose, which are long chains of glucose molecules, and proteins, which are long chains of amino acids.

A particular polymer, whether produced by nature or designed by scientists, has unique structural characteristics that determine its function and behavior. These characteristics depend upon the specific monomers involved, as well as how they are connected. For example, a polymer used to make fibers for clothing must be strong yet flexible (e.g., nylon), a polymer used to make pipes or flooring must be rigid (e.g., PVC), and a polymer used for making contact lenses must be hydrophilic enough to attract water molecules and allow lubrication of the eye.

Polymers typically are made by either (1) condensation reactions or (2) addition reactions. In an addition reaction, monomers containing C=C double bonds join end to end, the double bonds are replaced by C—C single bonds that connect the monomers, and no by-products are produced. In a condensation reaction, when two monomers join together, a by-product molecule. Polymers can be classified as copolymers or homopolymers. **Copolymers** are made of two (or more) different monomers that alternate in the chain, whereas **homopolymers** are made from identical monomers. Polymers also can be classified as chain-growth polymers or as step-growth polymers. **Chain-growth polymers** are made by chain reactions in which additional monomers are added to the end of a growing chain. **Step-growth polymers** are made when monomers with reactive functional groups at each end combine. Homopolymers usually are made by chain-growth with a single monomer, for example the formation of polyethylene from the ethylene monomer. Copolymers usually are formed by step-growth, such as the (repeated) reaction between the carboxylic acid and amide groups in 1,6-hexanediamene and sebacoyl chloride to form Nylon 6,12.

An example of an addition/chain-growth reaction is the formation of polyethylene:

$$\ldots + \begin{matrix} H & & H \\ & C{=}C & \\ H & & H \end{matrix} + \begin{matrix} H & & H \\ & C{=}C & \\ H & & H \end{matrix} + \ldots \rightarrow \ldots \begin{matrix} & H & & H & & H & & H & \\ & | & & | & & | & & | & \\ - & C & - & C & - & C & - & C & - \\ & | & & | & & | & & | & \\ & H & & H & & H & & H & \end{matrix}$$

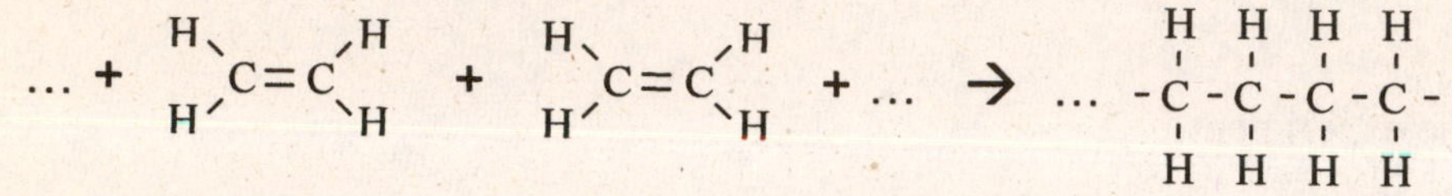

An example of a condensation/step-growth reaction is the formation of Nylon 6,12:

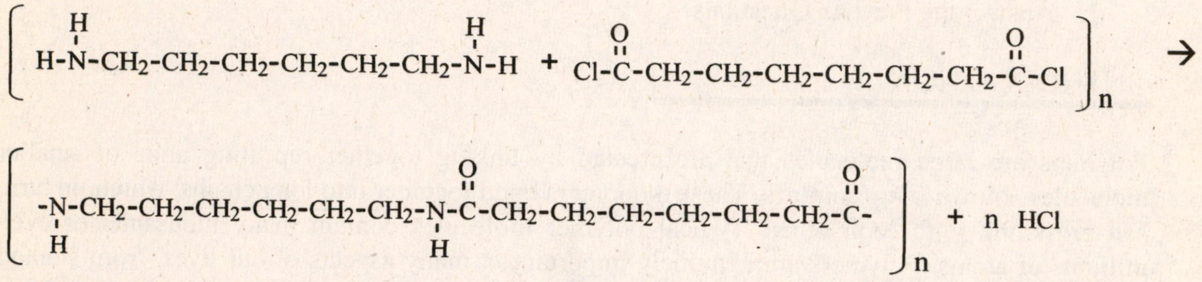

$$\left[\text{H-}\overset{\text{H}}{\overset{|}{\text{N}}}\text{-CH}_2\text{-CH}_2\text{-CH}_2\text{-CH}_2\text{-CH}_2\text{-CH}_2\text{-}\overset{\text{H}}{\overset{|}{\text{N}}}\text{-H} + \text{Cl-}\overset{\text{O}}{\overset{\|}{\text{C}}}\text{-CH}_2\text{-CH}_2\text{-CH}_2\text{-CH}_2\text{-CH}_2\text{-CH}_2\text{-}\overset{\text{O}}{\overset{\|}{\text{C}}}\text{-Cl} \right]_n \rightarrow$$

$$\left[\text{-}\underset{\text{H}}{\underset{|}{\text{N}}}\text{-CH}_2\text{-CH}_2\text{-CH}_2\text{-CH}_2\text{-CH}_2\text{-CH}_2\text{-}\underset{\text{H}}{\underset{|}{\text{N}}}\text{-}\overset{\text{O}}{\overset{\|}{\text{C}}}\text{-CH}_2\text{-CH}_2\text{-CH}_2\text{-CH}_2\text{-CH}_2\text{-CH}_2\text{-}\overset{\text{O}}{\overset{\|}{\text{C}}}\text{-} \right]_n + n\ \text{HCl}$$

The New Materials Division of UniChem International is working on the development of new polymers and new products that can be devised using them. You and your fellow research and development specialists have been given the materials needed to make a variety of polymers, determine their behavioral characteristics, and hopefully invent some different uses for them.

LEARNING OBJECTIVES

Be able to:

- Make different types of polymers.
- Classify polymers according to type.
- Draw structures of monomers and the polymers they form.
- Determine functions and uses of different polymers.

APPARATUS

Chemicals:

- Sodium polyacrylate
- Sodium chloride (NaCl)
- Borax
- White glue
- Food coloring
- 4% Polyvinyl alcohol solution
- 1,6-Hexanediamine solution
- Sebacoyl chloride solution
- Acetone-water mixture
- Toluene diisocyanate
- Poly(propylene oxide) glycol

Equipment:

- Teaspoon
- Small paper cups
- Wooden tongue depressor
- Quart plastic Ziploc bag

- Are your safety glasses on?
- Food coloring can stain your skin and clothing. It is recommended that you wear plastic gloves if you use it.
- Procedure 4 (making nylon) should be performed in a well-ventilated room to minimize breathing the vapors. Also, wear plastic gloves to handle the nylon "rope."
- The solutions used in Procedure 5 (polyurethane foam) are toxic. Wear plastic gloves and work in the fume hood.

Clean-up: 

- Follow your supervisor's directions.
- Wash your hands thoroughly with soap and water before you leave the laboratory.

PROCEDURE 1

Investigating Sodium Polyacrylate:

1. Add water to a 150-mL beaker until the water reaches the 100-mL line.
2. Then add a level teaspoon of sodium polyacrylate to the beaker, and mix thoroughly. Record your observations.
3. Next, add a level teaspoon of sodium chloride to the water–sodium polyacrylate mixture. Record your observations.

Refer to Procedure 1 section of your Report Sheet on p. 222.

PROCEDURE 2

Making GLUEP:

1. Add warm water (about 40–50°C) to a 250-mL beaker until the water reaches the 125-mL mark.
2. Then add one heaping teaspoon of borax to the warm water in the beaker, and mix well.
3. In a 600-mL beaker, add white glue to the 200-mL mark.
4. Then add water to the 600-mL beaker until it reaches the 400-mL mark, and stir to mix the glue and water thoroughly. (If you are feeling young at heart, you may add few drops of food coloring to the mixture as well.)
5. Finally, pour 100 mL of the borax solution into the glue-water solution in the 600-mL beaker, and mix well. (Save the remaining 25 mL of the borax solution for use in Procedure 3.) When the mixture is thick enough, dump it out of the beaker, and knead it well. Congratulations—you have made GLUEP!
6. Now that you have made it, it is time to investigate the properties of GLUEP. Some possible questions to try to answer are listed below. Use your creativity to come up with other questions to answer and investigations to perform.
   - Does it pour easily?
   - Is it flexible?
   - Does it stretch?
   - Can in be molded into a shape?
   - How does it respond to pressure?
   - Does it matter if the actions are performed quickly or slowly?

Refer to Procedure 2 section of your Report Sheet on p. 223.

PROCEDURE 3

Making SLIME:

1. Add 4% polyvinyl alcohol solution to a 150-mL beaker until it reaches the 50-mL mark.
2. Then add 10 mL of the borax solution (left over from Procedure 2) to the 4% polyvinyl alcohol solution and mix well. (Again, for aesthetic enjoyment, you may add a few drops of food coloring if you are so inclined.)

3. When the mixture is thick enough, dump it out of the beaker, and knead it well. Congratulations—you have made SLIME!
4. Now that you have made it, it is time to investigate the properties of SLIME. Some possible questions to try to answer are listed below. Use your creativity to come up with other questions to answer and investigations to perform.
   - Does it pour easily?
   - Is it flexible?
   - Does it stretch?
   - Can in be molded into a shape?
   - How does it respond to pressure?
   - Does it matter if the actions are performed quickly or slowly?

Refer to Procedure 3 section of your Report Sheet on p. 223.

PROCEDURE 4

Making Nylon:

1. Carefully pour about 5 mL of 1,6-hexanediamine solution into a 50-mL beaker.
2. Then carefully pour about 5 mL of sebacoyl chloride solution on top of the 1,6-hexanediamine solution so that the sebacoyl chloride solution forms a layer on top of the 1,6-hexanediamine solution, and the two solutions do not mix. (One of the most effective methods for doing this is to tip the 50-mL beaker slightly and carefully pour the sebacoyl chloride solution down the side to keep the two solutions from mixing.)
3. Use tweezers to reach through the upper solution to grasp the interface between the two solution layers. Carefully and steadily pull directly upward. You will be forming a hollow-cored nylon "rope" as you do this. Do not let the rope touch the sides of the beaker as you pull it upward.
4. As you continue pulling the nylon out of the solutions, have some of the other research and development specialists help you carefully to spread out the "rope" on a layer of paper towels so that it does not stick together. Try to make as long a continuous nylon "rope" as possible—keep going until one of the layers is used up or the rope breaks.
5. Rinse the nylon "rope" in 50 mL of water-acetone mixture and measure its length (this may require some untangling and/or estimating).
6. Take a small piece of the nylon "rope" and place it in a small beaker. Pour a small amount of acetone into the beaker (just enough to cover the nylon). Use a stirring rod to stir the nylon in the acetone. Does the nylon dissolve?
7. Grasp a piece of the nylon "rope" and pull gently. Did the nylon break? If not, pull a little harder. How much force was needed to break the nylon?

Refer to Procedure 4 section of your Report Sheet on p. 224.

PROCEDURE 5

Making Rigid Polyurethane Foam:

Reminder: Perform this procedure in a fume hood!

1. Use a graduated cylinder to add 25 mL of water to each of two small paper cups, and

mark the levels of the water surfaces on the outsides of the cups. Then empty and dry the calibrated cups.

2. Add toluene diisocyanate to the 25-mL line in one of the paper cups.
3. Add poly(propylene oxide) glycol to the 25-mL line in the other paper cup.
4. Use a wooden tongue depressor to pour and scrape all the less viscous solution into the other cup. Then use the tongue depressor to quickly and thoroughly mix the two solutions.
5. Immediately put the cup into a quart-size plastic bag and partially close the bag, leaving a small hole for gases to escape.
6. Although it takes about 24 hours for the polyurethane foam to reach its full strength, you can perform some tests after allowing the foam to set and cool for about 30 minutes. Break off a pea-sized piece of the foam. Describe its appearance (i.e., color, structure, etc.). How hard is it to compress the foam? How strong is it? Place the piece of foam in a small beaker, and add a small amount of acetone (just enough to cover the foam). If the foam collapses or becomes sticky, it is soluble in this solvent

Refer to Procedure 5 section of your Report Sheet on p. 225.

NAME (print) ______________________________ DATE ______________
LAST FIRST

LABORATORY SECTION ______ PARTNER(S) ______________________________

REPORT SHEET

PRE-LAB ASSIGNMENT

Pre-Lab Questions:

1. There are many important synthetic polymers that are used to make many of the substances we use in our lives. Please fill in the table below with the requested information for four synthetic polymers. (Note that two of the polymers have already been chosen for you, and the other two are your choice to make.)

Name of Polymer	Name and Formula of Monomer	Structure of Polymer (and Trade Names)	Uses of Polymer
Polyethylene			
Polyvinyl chloride			

2. Write the chemical equation for the reaction between sebacoyl chloride and 1,6-hexanediamine to make Nylon 6,12. Show all the structures involved.

NAME (print) ____ DATE ______________________
LAST FIRST

LABORATORY SECTION ______ PARTNER(S) ____

3. Is Nylon 6,12 a homopolymer or a copolymer? Briefly explain.

4. Is Nylon 6,12 a condensation or addition polymer? Briefly explain.

5. What is the meaning of the numbers 6 and 12 in the name Nylon 6,12?

PROCEDURE 1

1. Record your observations

   a. before the water was added to the sodium polyacrylate.

   b. after the water was added to the sodium polyacrylate.

   c. after the sodium chloride was added to the water-sodium polyacrylate mixture.

2. Suggest several consumer products in which sodium polyacrylate would be useful. Explain your reasoning.

NAME (print) ________________________________ DATE ______________
LAST FIRST

LABORATORY SECTION ______ PARTNER(S) ______________________________

PROCEDURE 2

1. Record your observations for the formation of GLUEP.

2. Record your observations for the behavioral characteristics of GLUEP.

3. Suggest several consumer products in which GLUEP would be useful. Explain your reasoning.

PROCEDURE 3

1. Record your observations for the formation of SLIME.

2. Record your observations for the behavioral characteristics of SLIME.

3. Suggest several consumer products in which SLIME would be useful. Explain your reasoning.

NAME (print) ______________________________ DATE ______________
LAST FIRST

LABORATORY SECTION ______ PARTNER(S) ______________________________

PROCEDURE 4

1. Record your observations for the formation of nylon.

2. Record your observations for the behavioral characteristics of nylon.

3. How long a piece of nylon "rope" did you make?

4. Is nylon soluble in water or in acetone?

5. How difficult was it to break the nylon rope?

6. Suggest several consumer products in which nylon would be useful. Will these products work well using individual strands of nylon? If not, explain what must be done.

NAME (print) ____ DATE ____________________
LAST FIRST

LABORATORY SECTION ______ PARTNER(S) ____

PROCEDURE 5

1. Record your observations for the formation of polyurethane foam.

2. Describe the structure, porosity, compression resistance, texture and general appearance of the polyurethane foam.

3. Is the polyurethane foam soluble in acetone?

4. Suggest several consumer products in which polyurethane foam would be useful. Explain your reasoning.

NAME (print) ______________________________ DATE ______________
LAST FIRST

LABORATORY SECTION ______ PARTNER(S) ______________________________

CONCULSIONS

Summarize your findings on the polymer syntheses and testing you did and the product development ideas you came up with. State your recommendations and the reasoning behind them in complete sentences. Be sure that your conclusion is complete and concise.

Appendix

Error Analysis in the Chemistry Laboratory

All measurements made in the chemistry laboratory have some degree of uncertainty associated with them. The goal of any chemist is to produce data that are reproducible. Therefore, it is important to determine how much a measured value differs or deviates from the known, literature or true value. The difference between the measured value and the known, literature or true value is called **experimental error**. Error in the laboratory can not be ignored or neglected because data that is not reproducible is useless.

It is important to clarify right away what is and what is not experimental error. The word **error** does not mean "mistake." Misreading the number on the balance, entering numbers incorrectly into your calculator, and mathematical errors are mistakes or blunders but not errors. Mistakes or blunders can be fixed, especially if you know about them or catch them. All experiments will have some degree of experimental error associated with them. Experimental error can be minimized but not totally eliminated. There are two types of experimental error in the chemistry laboratory: systematic and random.

Systematic error is due to a flaw in an instrument or in the design of the experiment. Systemic errors are made each time a measurement is made and result in all values to being shifted in a systematic way that leads to inaccurate measurements of the true value. An example of a systematic error is an instrument that is not calibrated properly. Systematic error have a definite magnitude and direction and usually can be detected and corrected.

Random error is error that fluctuates from one measured value to another. Random error usually is observed by making multiple trials of a particular measurement, such as measuring 2.00 g of sodium chloride. If you were to measure out 2.00 g of sodium chloride five times, you would notice that your measurements are not always the same. Random error displaces a measured value in an arbitrary direction and will affect the precision of the measurements. **Precision** is how closely the measured values are to each other. If all of the measured values are close together, the values have good precision. Random error is always present and is hard to correct. An example of random error is the reading of a burette. The graduations on a burette are usually measured in 0.1 mL increments. Common procedure for reading of a burette or any graduated glassware is to estimate the measured value to one decimal place of the smallest graduation on the measuring device. Therefore, the uncertainty associated with reading a burette is related to estimating the value that is in between two graduations, (e.g., 16.1 and 16.2 mL). The reading of a burette is subjective, and will vary with the person reading the burette and is an example of random error.

While most systematic error can be eliminated, random error is unavoidable. However, the goal of any chemistry experiment is to minimize random error and to eliminate systematic error. This will lead to data that have a high degree of precision and accuracy. **Accuracy** is defined as how close the measured value is to the true or literature value.

One of the most effective ways to estimate random error of a particular measured value it to make several trials of the measurement and calculate the mean or average and the **standard deviation,** which is the size of the uncertainty of the measurement. Both of these calculations are accomplished easily using a spreadsheet software package such as Excel® or can be accomplished with a calculator in statistical mode. If it is assumed that all systematic errors have been eliminated, the standard deviation of the measured values will allow us to infer the variability/precision of the measured values. The mean and the standard deviation for measured values should be recorded whenever possible. In some cases, it will be necessary to report the **percent relative uncertainty**, which compares the size of the standard deviation (or absolute uncertainty) with the size of the mean. The percent relative uncertainty can be calculated using the following equation:

$$\text{Percent relative uncertainty} = \frac{\text{standard deviation}}{\text{mean}} \times 100$$

In addition to reporting statistical error analysis, you also should cite possible sources of random error in the error section of your Report Sheet. The same random error will not be appropriate in every experiment, so you will need to think about the experimental procedure and the experimental techniques used to determine if any random error occurred. Also, it is important to address how the random sources of error may have affected your data.